Industrial Ecology: Policy Framework and Implementation

Braden R. Allenby
AT&T

Prentice Hall
Upper Saddle River, New Jersey 07458

Library of Congress Cataloging-in-Publication Data
Allenby. Braden R.
  Industrial ecology : policy framework and implementation / Braden R. Allenby.
     p. cm.
  Includes bibliographical references and index.
  ISBN 0-13-921180-2
  1. Industrial ecology.   I. Title.
  TS161.A38   1999
333.7--dc21   99-7300
                                                                              CIP

Acquisition Editor: Bill Stenquist
Editorial/Production: Ray Robinson - Douglas & Gayle
Editor-in-Chief: Marcia Horton
Assistant Vice President of Production and Manufacturing: David W. Riccardi
Managing Editor: Bayani Mendoza de Leon
Full Service/Manufacturing Coordinator: Donna Sullivan
Manufacturing Manager: Trudy Pisciotti
Creative Director: Jayne Conte
Cover Designer: Bruce Kenselaar
Editorial Assistant: Meg Weist
Compositor: Douglas & Gayle

©1999 by AT&T
Published by Prentice-Hall, Inc.
A Division of Simon & Schuster
Upper Saddle River, New Jersey 07458

Printed in the United States of America

10 9 8 7 6 5 4 3 2

ISBN 0-13-921180-2

Prentice-Hall International (UK) Limited, *London*
Prentice-Hall of Autralia Pty, Limited, *Sydney*
Prentice-Hall Canada Inc., *Toronto*
Prentice-Hall Hispanoamericana, S.A., *Mexico*
Prentice-Hall of India Private Limited, *New Dehli*
Prentice-Hall of Japan, Inc., *Tokyo*
Simon & Schuster Asia Pte. Ltd., *Singapore*
Editora Prentice-Hall do Brasil, Ltda., *Rio de Janeiro*

Contents

To my parents, Richard and Julia, who inspired me, and my family,
Carolyn, Richard and Kendra, who continue to.

Preface

For 25 years, "the environment" has been treated as both a new issue arising from the environmental activism of the early 1970s and a concern particularly associated with manufacturing and concomitant waste production and management. Driving this view has been locally evident environmental degradation—hazardous waste sites, polluted rivers, smog-laden cities around the world—and a concern for primarily human toxicity effects, especially cancer. Policy responses, predominantly in developed economies, have been shaped by these perceptions. Improvements in immediate environmental quality have occurred as a result, and few argue against the need for such measures (although many argue that implementation has been both economically and environmentally inefficient in many ways).

But scientific understanding and the structure of human economic activity have evolved considerably in 25 years, and policies and approaches that were accepted as adequate then are now obsolete. The relationship between fundamental natural systems and human cultural, technological, and economic systems is increasingly recognized as far more complex and profound than allowed for in the initial, naive policy responses. Environmental perturbations such as global climate change, loss of habitat and biodiversity, and degradation of soil, water, and air resources are not mitigated by localized clean-up mandates. Rather, they arise from intimate networks of connections between two sets of complex systems, those predominately natural and predominately human. Human and natural systems have co-evolved not for 25 years, or even in the centuries since the beginning of the Industrial Revolution, but for thousands of years. At this point, they are profoundly interwoven, even though we continue to treat them as separate systems for reasons of disciplinary boundaries, academic convenience, or psychology.

Moreover, the economies of developed countries are now predominantly service, rather than manufacturing or agriculture, oriented. In this light, a regulatory and policy structure focused on the environmental impacts of manufacturing is necessary, but is entirely inadequate as a comprehensive environmental management approach in a service economy. Environmental policies and regulation have begun to manage the impacts of the manufacturing activity; are barely beginning to address issues associated with the use of manufactured artifacts; and are blind to services.

It is not that current environmental policy and management approaches are wrong or should be discontinued. Rather, if humanity is to move beyond simply reducing a few obvious immediate risks to the much more fundamental effort to achieve an environmentally and economically sustainable future, much more must be done. It is the purpose of this text to explore the profound transition in policy, regulation, and management that will be required. This explication will be done in the context of the new and rapidly evolving multidisciplinary field of industrial ecology: "the science of sustainability."

Often before professors steeped in the powerful culture and learnings of their discipline, students understand that the reductionist, disciplinary approaches to environmental and sustainability issues they receive in other courses are powerful, but incomplete. The need for a comprehensive, multidisciplinary framework within which these complex issues may be understood in their totality generated the field of industrial ecology in the first instance, and now forms the rationale for this text. Although the book is primarily intended for advanced undergraduate and graduate students, particularly in the areas of engineering, economics, technology, business, environment, and policy, it is also written to be of use to those in industry, government, and non-governmental organizations (NGOs).

For most readers, industrial ecology and its policy implications are a new area, which can be fairly confusing if not presented properly. Accordingly, the first part of this book discusses the context within which policy, regulatory, and management decisions must be made. Here, introductory concepts such as sustainable development and sustainability, industrial ecology, the master equation, and the industrial ecology infrastructure are defined, and an industrial ecology intellectual framework suggested. Building on this understanding, the second part of the book begins with a discussion of perhaps the most conceptually simple but pragmatically challenging fundament for policy development in the industrial ecology and sustainability areas: that the systems to be managed are complex, and thus require a more sophisticated policy approach than is generally evident in practice. An overview of relevant legal and economic considerations is then provided, followed by a discussion of some of the implications of government and private firm structure for policy development and deployment. Finally, the third part of the text consists of four case studies that provide broad-ranging and concrete examples illustrating the principles and discussion elsewhere in the text.

It is somewhat difficult to write the first textbook in an area; precedents are useful not only as guides, but to codify the existing corpus of learning as well, ensuring that no major topics are slighted. Lacking this, I am particularly grateful to many people for the discussions and late night arguments that have been so important in helping me develop the concepts in these pages, including D. Allen, C. Andrews, J. Ausubel, J. Azar, D. Bendz, M. Chertow, N. Choucri, J. Ehrenfeld, P. Eisenberger, D. Esty, R. Frosch, T. J. Gilmartin, T. Graedel, B. Guile, K. Ishii, R. Laudise, R. Lifset, D. Marks, D. Rejeski, D. Richards, R. Schock, P. Sheng, R. Socolow, N. Themelis, V. Thomas, G. van Dijk, J. Watz, and I. Wernick (and with apologies to those I have inadvertently omitted). If there is wisdom in these pages, it is not mine alone; the mistakes, to the contrary, are mine. I also want to thank the President Emeritus of the National Academy of Engineering, Bob White, and the current President, Bill Wulf, for their support for the field of industrial ecology at a critical time, and Bill Merrell, President of the Heinz Center, for making industrial ecology a focus of the Center's work. Additionally, I thank my many friends and co-workers at AT&T, Lawrence Livermore National Laboratory, and the now-independent Lucent Technologies and NCR, from whom I have gained far more than contributed. Particular thanks go to my long-suffering secretary, Debbie Petrocy, who has not only had to struggle through many of the graphics, but keep a quasi-rational schedule for me as this book has been finalized. Finally, I would like to thank AT&T as an institution: both as a firm and working through such programs as the AT&T Foundation Industrial Ecology Fellowship grants, it has done much to support the evolution of the field of industrial ecology.

B. R. Allenby

PART I

A NEW POLICY FRAMEWORK

CHAPTER 1

Introduction

1.1 THE GLOBAL CONTEXT

The great Austrian economist Joseph A. Schumpeter once noted the "gale of creative destruction" that characterized capitalist systems. If this be true of capitalism in moderate times, today we are in the eye of a hurricane, shielded only by the illusory psychological assumption of stability that all humans carry, and the fact that the changes that are occurring are so fundamental, so complex, and so difficult for existing social structures to identify, that we, both as individuals and as societies, literally cannot comprehend them. It is against this background of accelerated and fundamental change in virtually every aspect of human endeavor—the revolutionary nature of the globalizing economy; the end of the bipolar Cold War geopolitical structure (almost caricatured as "the end of history"); the "information revolution"; the rise of the service economy; the devolution of power from the national state to local, regional, and global institutions and political interests, and transnational corporations and capital markets; the redefinition of virtually all social institutions, from the family, to churches, to universities, to private firms—that we must try to understand our subject matter.

This text presents an introduction to the policy issues arising from the integration of natural systems and human (artifactual) systems, a suite of issues commonly—if over-simplistically—referred to as "environmental concerns." These systems are, of course, co-evolved in fact: even the most urban system experiences weather, and must assure a supply of water, energy, and food, while the remotest biological communities bioaccumulate metals and chlorinated organics derived from human economic activity. What has yet to be developed, however, is the scientific understanding of what these linkages are and what they may mean in the future, not to mention the social, cultural, political, theological, and intellectual vision, understandings, and tools to deal constructively with what amounts to the beginning of a new chapter of human history.

This is, no doubt, somewhat daunting. Nonetheless, there are some unifying underlying themes that provide a context within which it is much easier to understand industrial ecology and its policy dimensions. Many of these themes will reappear throughout this book, and a brief introduction to them at this point will be useful in helping the reader organize his or her thoughts.

1. At this point in human history, a globalized economy and society is evolving that will not necessarily be homogenous. Rather, it will become more complex in the technical sense: There will be more communities, units, systems, interests, political and social entities, and technology clusters at many different levels, and concomitantly more relationships among them. Pressure on global and regional physical, biological, and chemical systems and cycles will require that impacts on these systems, which could previously be treated as externalities, will be internalized or cause major social and economic disruption. Understanding this complex and globally linked landscape using our current approaches will be difficult: We need to develop far more power in our systems models and thinking and much different and more sophisticated policy structures if we want to continue to constructively guide our future. Whether we can do so (and how rapidly) is an open question that this book attempts to address in a preliminary manner.

2. The earth is increasingly an engineered world in which human choice and technology determine the structure not only of human lives and environments, but for all life as a whole. It is now apparent that this trend is not new, but has been developing since the beginning of the Industrial Revolution, with its concomitant acceleration of human population and consumption levels, and linking of previously disparate local and regional patterns of human activity into globally synchronized systems. Indeed, in many ways, we already live in such a world; the principle reason we don't recognize this is that the engineering of the globe has not been planned. It has simply occurred without our conscious guidance. (This is, by the way, intended neither as technological optimism nor pessimism, nor is it presented as a normative "good" or "bad" outcome; it is a strong hypothesis based on historical data and easily projected trends.) A world population of 5.7 billion, moving toward 8 to 10 billion in the next 100 years, seeking material wealth according to current cultural norms in developed countries, will require such a world.

This strong hypothesis should not be taken, however, to imply that we have the appropriate technical, scientific, and policy tools to manage, or engineer, such a world at this point. The degree and kind of controls that such a complex, interlinked, and highly nonlinear system would imply are currently beyond the state of the art. Relevant parameters are neither well defined nor, in many cases, yet identified. Rather, this hypothosis points us toward what might become a new form of engineering—call it "earth systems engineering"—that, recognizing the complex and unpredictable nature of the interactions among artifacts, technological systems, culture, and the physical world, is appropriately humble, experimental, and decentralized.

3. Western concepts of reductionist science (study smaller and smaller units of a field in greater and greater depth) and linear, short-term, and obvious causality will have to be augmented (not replaced) by more systems-based, comprehensive approaches. This will be difficult because many familiar institutions—academic institutions, private firms, political structures—implicitly embody reductionist, discipline-oriented approaches, and are strongly resistant to integrative initiatives (as anyone who has worked with integrative centers in universities can attest). A further complication is that funding and research support systems (such as journals, conferences, and award and recognition programs) tend to be oriented toward disciplines and specialized fields, not integrative efforts. It is not over-simplifying by much to say that Western societies in general don't know how to do integrative science. An interesting question is whether Asian societies such as China and Japan, reflecting a more organic and holistic world view, can address integrative fields such as industrial ecology more effectively. If so, this raises the intriguing possibility that industrial ecology can be one of the grounds for a synthesis of the two great intellectual traditions of Eastern organic holism and Western reductionism.

4. Policy generally functions in the short term and focuses on the interests of a specific geographic area limited either by political structure (e.g., terms of office, national state boundaries), or by more fundamental human psychological constraints. Most people, for example, don't think beyond a time horizon of a few years and a geographic region that, at most, encompasses their country. Many of the natural and artifactual systems that are the subject matter of industrial ecology, however, have patterns that express themselves only over many decades, if not centuries, and over continental, if not global, geographical scales. A critical question, therefore, is how policy systems can be developed that integrate gracefully and robustly over very disparate temporal and spacial scales.

5. The degree to which virtually every modern institution—the national state, major religions, the private firm, universities and academic institutions, the family unit, the geographically localized community—is changing is both unparalleled and little recognized. This means that there are no firm points from which one can begin to develop intellectual frameworks and policy systems; it is all in rather fundamental flux. Cultural systems and humans tend to find such an environment uncomfortable, and will accordingly try to deny the reality of such all-encompassing change. Whether this is desirable is immaterial: It is a real phenomenon, and a difficult challenge for policy development.

6. While sustainability, and the specific form known as sustainable development, will be discussed at great length later in this section, it bears emphasizing here that the terms are not just ambiguous because of current lack of knowledge, but because they involve social choice. The relationship among human population levels, patterns of economic behavior (including choice of technologies and consumption patterns), and cultural systems and supporting natural systems is a complex one. In particular, it is highly likely that there are a number of states that could be sustained over some period of time measured in human generations, ranging from a Malthusian world where human population levels are maintained by the mortality rate, particularly among the poor, to a world that is consciously designed to support a high level of biodiversity and a limited human population. Intergenerational and intragenerational equity are thus not prerequisites of sustainability, at least in the short term, although they might be highly desirable.

Sustainability thus implies the existence of choice, although probably at a social, rather than individual, level. This, in turn, raises the question of "social free will," an interesting and difficult concept. Given the religious, social, cultural, political, economic, and other constraints that exist, how free is a society (or global society taken as a whole) to choose alternate paths? What does it mean to exercise (presumably bounded) free will at the level of the social organization, be it private firm or national state, rather than at the individual level? Must bounded social free will be assumed as a prerequisite for evolution of human activity toward sustainability, and, if so, what does the concept mean in practice?

The ambiguity of the concepts of "sustainability" or, more specifically, "sustainable development" does not mean, however, that it is of no value. Indeed, the term "sustainability" is used throughout this text as a useful and easily comprehended way to refer to a desirable and maintainable global economic and social structure. So long as the caveats are born in mind, this facilitates the presentation of material that is already complex enough.

7. Evolution toward an economically and environmentally efficient economy will differentially favor certain industrial sectors and technological systems and disfavor others. Sector definitions may well shift as economic activity rapidly evolves. For example, agricultural technology may become a critical process technology for pharmaceutical production. Understanding these trends will help national states and societies direct their research activities and resources, including employee skills, in the appropriate direction.

Among the sectors that will probably become more important are electronics and telecommunications, as the economy becomes more information dense and complex; biotechnology and ecological engineering, as biological systems and biomaterials are substituted for more energy- and material-intensive engineered systems; and agriculture, as renewables increasingly replace non-renewable resource streams. Among the sectors that will probably undergo significant change are the extractive industries, as material loops in the economy begin to operate more efficiently; transportation sectors, as energy prices rise to reflect internalization of rising social costs associated with energy production; and basic material industries such as steel and petrochemicals.

8. Environmental issues are occasionally framed in apocalyptic terms. It is highly unlikely, however, that human activity at anywhere near current levels threatens "the world," "life," or even the existence of the human species. What is threatened from a human perspective, rather, is the stability of global economic and social systems: Should they shift dramatically or even collapse, the impacts on human happiness, health, and mortality could be substantial. This is not to minimize the irreversible effects of, for example, loss of biodiversity and habitat that are clearly occurring even now, or the need for prompt development and deployment of ameliorative policies—which is, after all, the rationale for this text. Nonetheless, effective policy must be based so far as possible on clear goals and objective assessments of current and desired conditions, and the potential impacts of anthropogenic environmental perturbations should be evaluated in this light.

Finally, although it is not a theme, it is worthwhile to note another assumption that underlies the approach taken in this book: That, under optimal conditions, some kind of reasonable and acceptable path toward a desirable, environmentally and economically efficient, sustainable, global economy does exist. The available data do not assure us that this is the case. We may have already, for example, perturbed natural systems sufficiently to generate a discontinuous, and highly unpleasant, result such as rapid and substantial shifts in ocean circulation or climate patterns. Nor do we know that cultures and societies can change quickly enough to respond to challenges posed by the integration and globalization of natural systems and economic behavior, at least in the absence of a potentially disastrous crisis. A council of despair, however, is equally premature. More tellingly, it is also contemptible. If, by our actions now, we are likely able to increase our understanding of these profound issues, and develop and deploy policies that support a graceful evolution toward a sustainable world for our children and the life with which they share it, it is our ethical obligation to do so.

1.2   THE NEED FOR AN INDUSTRIAL ECOLOGY APPROACH

Change, and especially fundamental change, is never easy. It always imposes costs, regardless of the benefits that eventually accrue. It should thus be a fundamental rule that whenever a new paradigm is proposed, the need for it should be justified. Accordingly, we must begin our inquiry by determining whether the current situation suggests a need for a shift in paradigms to begin with.

The necessity for fundamentally changing the current approach to environmental issues arises because of the critical assumption on which current environmental policy is based: That environmental perturbations can be mitigated by treating them only as they are manifested without serious regard for the economic activity that caused them. In essence, this approach treats environmental considerations as if they are overhead, not strategic, for consumers, producers, and society itself. An *overhead effect* is one that is ancillary to the primary activity of an individual, a firm, or society, while a *strategic effect* is one that is viewed as integral to the primary activity. In this case, virtually all the relevant actors, including producers, consumers, and society as a whole, have operated as if environmental problems can be managed as overhead. One important indicator of this overhead status for environmental issues is the extent to which environmental policies and inputs have been separate, in almost every country and in international organizations, from other policy arenas perceived as more critical, such as technology policy, defense policy, economic policy, and trade policy.

Thus, environmental problems have been widely perceived as local in space and time, and are frequently associated with a single substance, such as the pesticide DDT, polychlorinated biphenyls (PCBs), or heavy metals. Social responses are still ad hoc, tend to focus treatment on single media, and reflect a strong bias toward remediation of existing localized problems—individual air- or watersheds or specific waste sites—that in many cases pose little, or easily managed, risk. Compliance activities, especially in the United States, are usually focused on limiting emissions of undesirable materials through the use of emission control technologies—the so-called "end-of-pipe" approach. Even pollution prevention programs generally focus on relatively simple adjustments of existing production technologies, and virtually all of them focus on manufacturing, which is only one life-cycle stage in a product's passage through the economy, and frequently not the one generating the most significant environmental impact. Pollution prevention in the manufacture of automobiles, for example, is desirable but clearly does not begin to address the major environmental implications of that technology system.

The existing paradigm, therefore, can be fairly characterized as one within which environmental issues are seen as overhead, ancillary to more important and fundamental economic activity. In this case, as in many, however, the issue is not whether the current paradigm of environmental regulation is working: As implemented in most developed countries, it has demonstrably resulted in cleaner air and water and less toxic loading of the environment. The question, rather, is whether the paradigm is adequate to explain the new information and data that have accumulated since it was first developed, or whether it must, in turn, be subsumed into a broader approach. The answer to this latter question is clearly yes.

The prevailing "overhead paradigm" and the evolving industrial ecology paradigm are compared in Table 1.1, which reflects the acceptance of the need to integrate

TABLE 1.1   Traditional Environmental Policy versus Industrial Ecology

Primary Activity	Goals	Focus of Activity	Relation of Environment to Economic Activity	Disciplinary Approach	Assumed Nature of System
Remediation and emissions controls	Reduce local and immediate human risk: clean air and water	Individual site, media, or substance	Overhead/ Externality	Environmental professional, reductionist	Simple
Industrial Ecology	Global sustainability, including mitigation of global climate change, loss of habitat and biodiversity, and degradation of water, soil and atmospheric resources	Materials, products, services and operations over life cycle	Strategic and integral	Physical sciences; biological sciences; social sciences; law and economics; technology and engineering; highly integrative	Complex

environmental, technological, and economic activity globally. This is best reflected in the policy endpoints. Remediation and compliance aim at the reduction of localized risks, often defined only in terms of human risk. The endpoint of the industrial ecology approach, on the other hand, is a globally sustainable economy. The ad hoc focus on symptoms of unsustainable economic activity—specific media impacts or waste sites, for example—is augmented and made more efficient, not replaced, by a far more comprehensive approach that focuses on production and consumption patterns throughout the economy.

This shift in viewpoint is significant for two reasons. First, while reducing local risk is clearly desirable, it will never be enough in itself to assure sustainability. The environmental perturbations threatening sustainability—such as loss of biodiversity, global climate change, stratospheric ozone depletion, and global degradation of water, soil, and air resources—are simply not addressed by remediation or localized compliance programs. Mitigating them requires fundamental changes in technology and in economic and cultural behavior, not just the establishment of a fund to support clean-ups.

The automobile provides an illustration. In recent years, the environmental performance of each automobile, measured in emissions per mile, has vastly improved. From 1968, when controls were first imposed, to 1992, average vehicle emissions of volatile organic compounds (VOCs) and carbon monoxide decreased by 96 percent. The first controls on nitrogen oxides (NOx) emissions were implemented in 1972, and by 1992 had resulted in reductions of emissions in new cars by 76 percent. These environmental improvements were achieved even while performance of the technology, both in terms of resource efficiency (miles per gallon and horsepower per unit engine size) and of acceleration, was improving simultaneously.

The automobile's emissions reductions can be attributed to compliance programs—and compared to other compliance programs, relatively sophisticated ones at that—in the sense that specific control technologies were not mandated a priori, and they were focused on product use, not just product manufacture. On the other hand, these compliance programs clearly don't begin to address the major environmental impacts of automobiles. They don't address culture: Much of the environmental gain per vehicle attributable to better technology has been undermined by the fact that there are now more automobiles per capita in developed countries, each driven further per year, and that customer choice in places like the United States and Europe is shifting toward so-called sports-utility vehicles and trucks, which are far less fuel efficient than mid-size sedans. They also don't begin to address the major environmental implications of the infrastructure required by the automobile: The asphalt, aggregate, concrete, steel, and other materials required for roads, bridges, parking lots, malls, and other built infrastructure, and the concomitant impacts on local ecosystems of construction activity. Nor are the environmental impacts attributable to the petroleum industry addressed, which in large part exists to fuel the automobile or to supply infrastructure materials (e.g. asphalt). Most critically, the role of the automobile in creating patterns of geographical population dispersion and behavior, which in turn create and maintain high demand for personal transport—thus locking societies such as the United States into high automobile use patterns—is not even considered.

As this example suggests, compliance programs directed toward specific emissions or substances can be valuable. They are, however, inevitably and profoundly non-systemic: The more complex the system under consideration, the less useful such an approach generally is. Accordingly, an industrial ecology approach, which, for example views the automobile and its supporting infrastructure as a complex technological system, is a necessity if the goal is evolution toward a sustainable society. Moreover, failure to implement most remediation or compliance activities may increase local risk, but will seldom threaten the global environment (there are some exceptions: controlling emissions of CFCs is essential to limiting the risks of stratospheric ozone depletion). Failure to implement policies based on industrial ecology can, however, result in irreversible global environmental degradation.

The automobile example makes another point as well. Generally, remediation and compliance programs assume a complete understanding of the systems involved, so that a specific regulation can be targeted to have the desirable effects without causing any unanticipated side effects elsewhere. If properly implemented, this is an appropriate approach to specific, well-defined, localized hazards—a chemical spill, for example. This assumption is valid only for very simple systems, however, and becomes less appropriate the more complex the system becomes.

For example, some people have suggested requiring that bismuth-based solders replace lead-based solders in electronics manufacturing, given the apparent difference in toxicity between the two materials. Disregarding the fact that the toxicity of bismuth has not been well characterized, especially for non-human organisms, this apparently justifiable suggestion raises major problems as soon as a systems approach is adopted. For example, bismuth is generally produced as a byproduct of lead mining, so a major increase in demand for bismuth might actually increase the amount of lead ore mined. Second, during end-of-life secondary smelting of shredded electronics products, bismuth can "poison" copper, that is, render it unusable for subsequent electronics applications. Thus, in this case, an apparently simple and unquestionably beneficial targeted intervention (replace a more toxic material with a less toxic one) becomes problematic when the relevant system is considered in its entirety.

In fact, it is generally the case that regardless of the specific methodology used, assessments of environmental impacts of technologies or materials over their life cycle tend to reveal significant surprises. One example is the issue of paper versus plastic drinking cups, which was a fairly contentious debate until a life-cycle study by the Government of the Netherlands surprised everybody by demonstrating that, under most circumstances, both paper and plastic disposable cups are preferable to ceramic ones, a quite counterintuitive result. Another life-cycle study indicated that, from an environmental perspective, it is probably better to discard polyester blouses after one wearing and buy new ones, rather than wash and wear them again (because of the energy and material inputs required by the washing/drying process).

The point is not that traditional remediation and compliance activities, when properly understood and applied, are not desirable. Rather, it is that the integration of science, technology, and environment—the study of industrial ecology—involves complex systems and interactions that, in many cases, are not well understood or characterized. Some of the differences between simple systems and complex systems are discussed in Chapter 9; at this point, it is only necessary to recognize that a goal of sustainability will require very different kinds of policies than those traditionally relied on in the past.

It is premature to claim that the policy implications of the industrial ecology paradigm are understood. Indeed, in many cases it is apparent that both the data and the understanding to even ask the appropriate questions are lacking. What is clearly apparent, however, is a fundamental paradigm shift is required to move beyond the initial social responses to immediate environmental challenges that have characterized the past 25 years to the far more difficult task of creating a sustainable world, that is, an environmentally and economically efficient world that can be maintained in a desirable state over an appropriate time period.

Because the intellectual underpinnings necessary to this task are not widely understood, Part I of this text begins with a description of the new intellectual framework that will support this journey. In Part II, the major policy implications of this new paradigm are discussed. Given the breadth and rapidly evolving nature of industrial ecology, it is difficult to understand the new paradigm from principles alone; accordingly, Part III contains four case studies, which have been chosen to provide illustrations of important industrial ecology dimensions. The first illustrates a qualitative, structured analytical tool for performing an industrial ecology assessment (an example of a "Design for Environment" or DFE tool); it serves as evidence that workable, practical tools to address such complex systems can, in fact, be developed. The second, more conceptual case study investigates the implications of a sustainable economy for a major stakeholder in the process, the private corporation. The third discusses the policy structure established by the Netherlands in this area, considered by many experts to be the most sophisticated in the world; it demonstrates the viability of integrated, national industrial ecology policy structures. The fourth case study investigates the integration of national security and environmental policy structures into a new "environmental security" approach. This is not only substantively relevant, but provides a useful illustration of the process by which issues previously regarded as overhead become strategic for society.

REFERENCES

Allenby, B. R., and D. J. Richards, eds. *The Greening of Industrial Ecosystems*. Washington, DC, National Academy Press: 1994.

Graedel, T. E., and B. R. Allenby. *Industrial Ecology*. Upper Saddle River, NJ, Prentice-Hall: 1995.

Graedel, T. E., and B. R. Allenby. *Industrial Ecology and the Automobile*. Upper Saddle River, NJ, Prentice-Hall: 1997.

MacKenzie, J. J. *The Keys to the Car*. Washington, DC, World Resources Institute: 1994.

Van Eijk, J., J. W. Nieuwenhuis, C. W. Post, and J. H. de Zeeuw. *Reusable Versus Disposable: A Comparison of the Environmental Impact of Polystyrene, Paper/Cardboard, and Porcelain Crockery*. Deventer, The Netherlands, Ministry of Housing, Physical Planning and Environment: 1992.

EXERCISES

1. Monitor your consumption of goods, including food, for a week.

   a. In what ways do you as a consumer treat environmental concerns as integral to your consumption decisions?

   b. Do you have enough information to know how to take environmental considerations into account in your consumption decisions? What additional information do you need?

   c. Assume you have enough information to make environmentally appropriate consumption choices. To what extent can you, acting independently, modify your consumption choices, and to what extent are they constrained by broader economic, technological, and cultural factors embedded in the products and services available?

2. Provide at least four reasons why you think people buy environmentally and economically inefficient vehicles for personal use.

   a. In your opinion are the primary reasons for this behavior: technical, cultural, psychological, or economic?

   b. Assume that you are in a position to encourage or mandate a change in this behavior. Should you? Why or why not? What policy tools would you use if you chose to do so?

CHAPTER 2

Overview of the Industrial Ecology Intellectual Framework

It is one thing to recognize that the past approach to environmental issues, with its focus on remediation and compliance, is necessary, but is in itself inadequate to approach and maintain a sustainable world. It is another to provide the intellectual framework that can support initial efforts at all levels, from the theoretical to the practical, to define and expand the industrial ecology paradigm that must replace it. Such a framework is presented here, and is explained in more detail in subsequent sections. It is noteworthy that, although each level of the hierarchy requires a different set of implementing policies, these policies must be developed so that they do not conflict, but rather reinforce one another. This will occur only by happenstance in the absence of a unified, internally consistent, framework.

Figure 2.1 presents the intellectual framework in outline. The highest level represents the vision, which is captured in the concept of *sustainable development*. This was defined in 1987 by the Brundtland Commission, which originated the term, as "development that meets the needs of the present without compromising the ability of future generations to meet their own needs." It is a somewhat value-laden term, implying for many people, for example, redistributing wealth intergenerationally (from ourselves to future generations) and intragenerationally (from wealthy developed nations to developing nations, and from wealthy elites to the poor). It also implies for some a need to aggressively restrict population growth, an activity that, regardless of its environmental benefits, is highly controversial.

It is also a very ambiguous term. To begin with, existing data cannot prove or disprove the contention that, in fact, present economic development can continue without compromising the ability of future generations to meet their own needs. Equally as important, it is not clear what sustainable development might mean operationally. For example, it is obviously meaningless to consider small subsystems sustainable, when sustainable can only be defined in terms of the global system itself. Thus, for example, a firm that chooses to call itself sustainable may mean that it has adopted sustainability as its vision, but it cannot be sustainable in actuality so long as it is integrated into an unsustainable global economy. It would be equally implausible to call a product or process sustainable at this point in the evolution of the Industrial Revolution. One could not make these assertions of sustainability without understanding the global

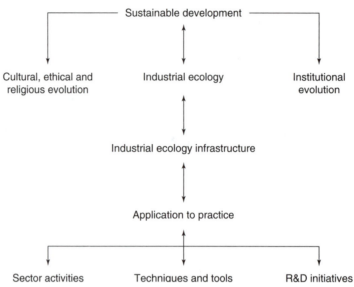

Industrial Ecology Intellectual Framework

FIGURE 2.1   Although simple, the industrial ecology intellectual framework provides a useful structure for understanding the components of sustainable development.

human and natural systems as an integrated whole, and being able to project their stability through time, which is clearly not the case now and will not be for the foreseeable future.

This does not mean, however, that sustainable development cannot provide the general goal toward which environmentally preferable activities can be directed. For example, the Netherlands has adopted the goal of becoming sustainable within one generation. This is not because they believe that as a small, heavily exporting nation that is only a part of regional European watersheds and airsheds they share with Germany, France, and others, they can be sustainable in a global sense. Rather, defining their goal in such a manner has enabled them to do path-breaking research into what sustainability might mean for a modern, developed country, and what metrics might be appropriate for determining progress towards those goals. Similarly, the Institute of Electrical and Electronics Engineers, Inc. (the IEEE), the world's largest technical professional society, considers sustainable development a worthy vision, while noting in its White Paper on Sustainable Development and Industrial Ecology (1995) that "standing alone, [sustainable development] . . . cannot guide either technology development or policy formulation."

The second level of the hierarchy, *industrial ecology,* is the multidisciplinary study of industrial systems and economic activities, and their links to fundamental natural systems. It provides the theoretical basis and objective understanding upon which reasoned improvement of current practices can be based. Important disciplines contributing to industrial ecology include the physical and biological sciences, engineering, economics, law, anthropology, policy studies, and business studies. Even given this broad scope, however, it is important to note the distinction between the vision of sustainable development, which is heavily normative and thus relies on political and cultural systems for its definition, and industrial ecology, which is an objective field of study, and thus relies on traditional scientific, engineering, and other disciplinary research for its development. Obviously, policy development and deployment will reflect these differences.

As Figure 2.1 illustrates, industrial ecology in itself is almost certainly inadequate to support the achievement of a sustainable global system; in other words, science and technology taken alone are not sufficient. Significant cultural, ethical, and religious evolution are difficult, but also necessary. For example, in many societies and for some major religious organizations, equal rights and professional opportunities for women are problematic. Yet many have argued that maintenance of reasonable population levels, not to mention intragenerational equality, may well depend on achieving such equality of opportunity.

Another equally important capability will be institutional evolution. As noted in Chapter 1, the rate of such evolution has rapidly increased since the end of the Cold War, mostly for reasons not associated with sustainability or industrial ecology. How these existing patterns of change will shift in an increasingly environmentally constrained world is not clear, even as the need for such adaptation becomes more apparent.

The third level of the hierarchy, the *industrial ecology infrastructure,* constitutes society's response to this question: "Assuming that private and public firms and consumers can be encouraged to behave in environmentally appropriate ways, what must the state and society in general provide so that they may do so?" It thus includes developing and implementing the legal, economic, and other incentive systems by which desirable behavior can be promoted, as well as the methodologies, tools, data, and information resources necessary to define and support such behavior. One example might be the development of legal structures that promote environmentally appropriate behavior, and the development and diffusion throughout the global economy of environmentally preferable technologies. Examples of such policies might be environmentally sensitive government procurement regulations, military specifications, and military standards; the removal of environmentally and economically inefficient subsidies for virgin, as opposed to recycled, materials; and the removal of energy, transport, agricultural, fishery, and forestry subsidies, which distort production in those sectors in environmentally inappropriate ways.

The difference between these two levels can be illustrated by considering the need to understand the environmental and energy costs embedded in various materials in common applications, so that, where cost, technology, and other constraints permit a choice, the material with the least social cost—the smallest negative impact on society—can be selected (Figure 2.2). Obviously, this is an important component of better environmental performance; it is at present unfortunately true, however, that there is virtually no information on the integrated "social costs" of various materials available. Thus, for example, engineers on product or process design teams and others have no way of knowing what material option is preferable from a life-cycle environmental cost perspective. Experience as well as the preceding discussion about sustainability both indicate that the concept of "sustainable material use," while a useful goal, is far beyond the current state of knowledge.

Industrial Ecology Intellectual Framework: Materials Example

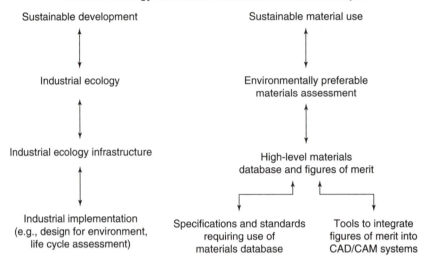

FIGURE 2.2 This figure demonstrates how the framework of Figure 2.1 is applied to a particular component of industrial ecology, in this case, material use. Similar frameworks can be constructed for energy, or for sectors such as forestry, agriculture, or fisheries.

This does not mean, however, that progress cannot be made at lower levels of the hierarchy. At a relatively aggregated level, it is necessary to develop the scientific and methodological basis for understanding the life-cycle environmental impacts associated with particular applications of different materials. Developing the data and methods to support an understanding of environmentally preferable material selection is an industrial ecology research issue. The need to take the next step is equally critical, albeit frequently overlooked: Translating that information into a form that is useful for designers and others who will actually embed the knowledge into the economic system. Practical tools at an appropriately aggregated level, and suitable for use in existing systems and industrial cultures, must also be built. In this case, it would mean ensuring that the data on environmental impact of various materials in particular applications, which in many cases might well be voluminous, uncertain, and internally inconsistent, are consolidated into a form that can easily be accessed by industry (for example, figures of merit that can easily be mechanically fed into automated CAD/CAM, or computer aided design/computer aided manufacturing, systems). This latter step would be an industrial ecology infrastructure project. The actual application of the knowledge about materials thus gained and codified are applied at the final level of the hierarchy.

Unlike the first three levels, the fourth level of the hierarchy, *application to practice,* is not primarily concerned with policies, but with implementation. While the specific activities undertaken will be different for different firms, different consumers, and different economic sectors, it will for all of them represent the level of immediate action based on industrial ecology principles as currently understood and translated into policy.

It is useful, however, to break this into three categories conceptually, even if there is some degree of overlap. The first represents those improvements that can be made in various sectors in the short term given the current state of knowledge. These amount to essentially incremental changes in existing economic practices, and in most cases are already under way, at least among leading firms. This does not mean that such changes cannot result in dramatic environmental improvement, however, or that they are not politically and culturally challenging, as any who have tried to implement such practices in private industry can testify.

The second category includes the techniques and tools, such as methodologies, data sets, rules of thumb, guides, standards, and procedures, that need to be developed to form the basis of a broader social capability in this area. It differs from the first category in the universality of applicability of the technique. For example, Design for Environment methodologies are included in the first category because their development and deployment tends to be limited to the manufacturing sector, particularly electronics and automotive manufacturing. Accounting systems that break environmental expenditures and liabilities out of overhead—so-called *green accounting systems*—are included in the second category, however, because such accounting methodologies can be productively used by private firms throughout the economy, regardless of sector.

The third category, research and development initiatives, includes those R&D activities that, although they can be initiated immediately, do not directly result in environmentally preferable practices. Rather, they contribute to the development of the knowledge base necessary to support the development of longer term, more fundamental applications. Unlike the sector applications, there may be little incentive for individuals or private firms to undertake these activities, either because they are public goods; or because of the riskiness and minimal current value (because any return would occur only in the far future) of any possible return. In many cases, therefore, these initiatives are likely to require government or nonprofit support.

REFERENCES

Allenby, B. R. "Industrial ecology gets down to earth." *Circuits and Devices* (January 1994): 24–28.

Allenby, B. R. "A design for environment methodology for evaluating materials." *Total Quality Environmental Management* 5(4): 69–84.

Allenby, B. R. and D. J. Richards, eds. *The Greening of Industrial Ecosystems.* Washington, DC, National Academy Press: 1994.

Graedel, T. E. and B. R. Allenby. *Industrial Ecology.* Upper Saddle River, NJ, Prentice-Hall: 1995.

Institute of Electrical and Electronics Engineers, Inc (IEEE), Environment, Health and Safety Committee, White Paper on Sustainable Development and Industrial Ecology, Washington, DC, 1994.

Socolow, R., C. Andrews, F. Berkhout, and V. Thomas, eds. *Industrial Ecology and Global Change.* Cambridge, Cambridge University Press: 1994.

Turner, B. L. II, W. C. Clark, R. W. Kates, J. F. Richards, J. T. Mathews, and W. B. Meyer, eds. *The Earth as Transformed by Human Action.* Cambridge, Cambridge University Press: 1990.

World Commission on Environment and Development (also known as the Brundtland Commission). *Our Common Future.* Oxford, Oxford University Press: 1987.

EXERCISES

1. Rank the following organizations in order of their capability to: a) generate a change in practices at the local level; b) generate a change in practices at the national level; c) influence people's attitudes at the local level; and d) influence the behavior and attitudes of their culture.

    1. A small, activist local environmental group.
    2. A small captive supplier to an automobile manufacturer.
    3. A large transnational manufacturer of electronics devices.
    4. A chemical workers union.
    5. A large technical research university.
    6. An elementary school.
    7. A national Minister of Science and Technology.
    8. A state Environmental Regulatory Agency.

2. Which of the preceding groups would you expect to be most interested in a series of releases of toxic gases in several states? In new data regarding the human health effects of a common industrial chemical? In global loss of biodiversity? In encouraging the implementation of industrial ecology? Why?

3. Some firms and nations have established themselves a goal of becoming sustainable in their operations. What would such a goal mean to you? Is this a useful or desirable goal? Defend your answer.

CHAPTER 3

Sustainable Development

3.1 SUSTAINABLE DEVELOPMENT: DEFINITION

The classic definition of sustainable development was provided by The World Commission on Environment and Development (also known as the Brundtland Commission) in 1987 in *Our Common Future* (43): "Sustainable development is development that meets the needs of the present without compromising the ability of future generations to meet their own needs." While this sentence is often cited, it is useful to understand the political context within which the Commission, at least, viewed the concept:

It [sustainable development] contains within it two key concepts:

- The concept of 'needs', in particular the essential needs of the world's poor, to which overriding priority should be given; and
- the idea of limitations imposed by the state of technology and social organization on the environment's ability to meet present and future needs.

Thus the goals of economic and social development must be defined in terms of sustainability in all countries—developed or developing, market-oriented or centrally planned. . . .

Even the narrow notion of physical sustainability implies a concern for social equity between generations, a concern that must logically be extended to equity within each generation.

Elsewhere, the Commission notes that (56), "sustainable development can be pursued more easily when population size is stabilized at a level consistent with the productive capacity of the ecosystem," and that "the challenge now is to quickly lower population growth rates, especially in regions such as Africa, where these rates are increasing." This leads the Commission to note that, "increased access to family planning services is itself a form of social development that allows couples, and women in particular, the right to self-determination."

3.2 IMPLICATIONS OF SUSTAINABLE DEVELOPMENT

There are some important initial points to be made about the concept of "sustainable development" as thus presented and commonly interpreted. Most importantly, the concept is anthropocentric, that is, it is focused on the *human* species. Other life is considered as it contributes to this end. The ethical questions about the value of species other than human are thus finessed. On the other hand, this formulation of sustainability clearly takes firm ethical positions on social issues: in particular, inter- and intragenerational equity is made a fundamental prerequisite for a state of sustainable development. Moreover, the needs for population control and equal rights and opportunities for women are clearly stated; these are religiously and culturally controversial as well.

Nonetheless, the concept of sustainable development as thus laid out is still ambiguous and impossible to operationalize: Given the current state of knowledge, no one knows what a sustainable global economy would look like. Moreover, the impact of the WCED's work, and the broad publicity it received, have had the unfortunate tendency to retard the debate about other possible sustainable states and the need for choices among the universe of potential sustainable futures. Sustainable development is one possibility among many, and it may well not be the most probable. After all, equality has yet to be achieved through human history, which suggests that it may be unlikely in the near future.

Additionally, the social rather than scientific approach embodied in the traditional definition of sustainable development has led to confusion about the systemic basis of sustainability. In particular, it has hindered the understanding that the concept of a sustainable subsystem in an unsustainable global system is fundamentally oxymoronic, even if the subsystem is equalitarian. Concepts such as "sustainable community," "sustainable firm," and "sustainable product," therefore, must be seen as generic indications of goodwill toward environmental issues, rather than indications that the entity in fact understands sustainability or has achieved such a state. At best, it means the achievement of increased environmental efficiency: a desirable goal, but not more.

More fundamentally, sustainability must be a characteristic of the global system as a whole, including both human activity in its totality and the underlying natural biological, chemical, and physical systems at all relevant scales. In a sense, the sustainability of any subsystem of the global system—be it a state, a firm, a region, or even an individual—can only be defined in terms of a sustainable global system, and cannot be meaningfully said to exist in the absence of its links to the greater whole. Moreover, sustainability is probably what students of complex systems would call "an emergent characteristic" of an appropriately organized global system, which means, among other things, that one cannot know that one is sustainable until, in fact, such a state has already been achieved. Additionally, if sustainability is anything, it is a process, not an endpoint: As human culture, technologies, population levels, and natural systems evolve, so will the parameters of sustainability. Accordingly, sustainability has a temporal component: sustainable for how long and within what time frame? It is not difficult at all, for example, to imagine a world that slips in and out of desirable states that are each sustainable over some time frame, a sort of "punctuated sustainability."

Having said this, however, it is also apparent that sustainable development as a vision can be an important guide to significant conceptual progress. Thus, the Netherlands established a goal of being sustainable in one generation not because they really believed a small country whose watersheds and airsheds for the most part lie beyond its borders, and whose economy is heavily import and export oriented, could really be sustainable on its own. Rather, such a vision gave them a means of beginning to define what moving toward a sustainable state might mean: what metrics might be useful; what policies might support progress in the right direction; what risks, costs, and benefits might be involved. Used in this way, the concept can generate significant insights.

This discussion leads to an important aspect of sustainable development and sustainability generally: It is both a science and technology challenge and a social science challenge. It involves both objective research and significant political and ideological issues. While these two dimensions are inextricably linked in policy development (Figure 3.1), they are conceptually distinct. It is desirable to keep them this way, as it separates out questions of fact and data, which can be resolved by further research, from the inevitably contentious political debates over the desirability of social changes that progress toward sustainability may well require. The latter are not susceptible to technological resolution, but require the evolution of social values. Confusing the two serves little purpose. Natural systems will not change their behavior because it is deemed politically inappropriate; on the other hand, a vigorous political debate when social change is perceived as necessary is both appropriate and desirable. Only thus can a stable social consensus offering some probability of being robust over time be achieved.

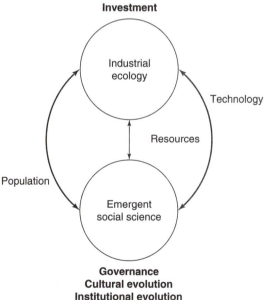

Policy Development Framework

FIGURE 3.1   It is important to recognize that social science, as well as engineering and physical science, will be co-evolving, coupled systems.

3.3   IDEOLOGICAL DIMENSIONS OF SUSTAINABLE DEVELOPMENT

The concept of sustainable development as defined by the WCED raises at least four socially and culturally contentious issues. First, it implies some limitation of growth in physical human activity, in particular in the velocity and magnitude of use and disposal of materials, and in the accumulation of physical stock. Note, however, that it does not imply a termination of development or evolution of human institutions or the global economy itself. Nonetheless, especially for some of those trained in neo-classical economics, this assumption is problematic.

Second, it implies a limitation to unfettered human reproduction. This is problematic for many powerful pronatalist institutions. For example, Dr. Navarro-valls, the Director of the Catholic Church's Holy See Press Office in Rome, wrote concerning a conference on population control held in Cairo in 1994:

> There is said to be, however much disputed on empirical grounds, a world population crisis. In this doubtful view, the need to control populations becomes the paramount ethical and political issue. [There is no need to] control human population according to a questionable theory of world resources and human needs. . . . Civilization is at stake [if population control is accepted]. We would be foolish to see in the Cairo Conference anything less.

A related third issue is that sustainable development implies equal status and equal rights for women. There are many religious groups, including Hindu, Islamic, and Christian ones, for which this would represent a fundamental and difficult change.

Fourth, the idea of redistributing wealth both intragenerationally and intergenerationally tends to be opposed by those who would lose assets in the process. Intergenerational equity involves questions of resource consumption and preservation of current "global assets" such as biodiversity, so that current consumption does not inequitably impact future generations; it is equity over time.

Intragenerational equity not only involves the question of resource shifts between developed and developing states, but also within states as well; it is equity over space. Many developing countries, for example, are characterized by elites that appropriate much of the resources of the country: An example is the way that Nigerian rulers have, over past decades, maintained control over huge revenues from oil production, while the vast majority of citizens of that country remain poor.

Recall that it is not clear that equity of either kind is necessary, as opposed to perhaps desirable, for a physically sustainable world. This is perhaps hopeful, if unfortunate, as the history of the species does not reveal any period where equality on such a scale was the norm. The historical record would tend to argue that, if sustainability has to wait on equity, the chances of achieving the former would be considerably diminished. In short, we as a species may achieve sustainability far sooner than we achieve sustainable development.

Another important dimension of sustainable development and sustainability in general is created by the fact that individuals have greatly differing mental models concerning the stability of natural systems. These, in turn, significantly affect the ideologies people are inclined to in this area. The two extremes and the centrist position are illustrated in Figure 3.2. There are those who believe that nature as a whole is virtually immune to human perturbation; no matter how many people there are or what technologies are used, natural systems are inherently stable (Figure 3.2 a). This ideol-

Mental Models of Nature

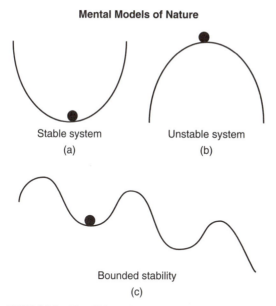

Stable system
(a)

Unstable system
(b)

Bounded stability
(c)

FIGURE 3.2   The different images people have of the stability of natural systems are frequently unconscious and unquestioned. They can significantly affect how different people interpret data on environmental issues, however.

TABLE 3.1   Options for Technology-Society Interactions

Approach	Effect on Technology	Implications
Radical ecology	Return to low technology	Unmanaged population crash: economic, technological, and cultural disruption
Deep ecology	Appropriate technology, 'low-tech' where possible	Lower population, substantial adjustments to economic, technological, and cultural status quo
Industrial ecology	Reliance on technological evolution within environmental constraints: no bias for 'low-tech' unless environmentally preferable	Moderately higher population, substantial adjustments to economic, technological, and cultural status quo
Continuation of status quo	Ad hoc adoption of specific mandates (e.g., CFC ban): little effect on overall trends	Unmanaged population crash: economic, technological, and cultural disruption

ogy corresponds to a strong belief in continuing the status quo (Table 3.1), and a relative lack of concern for population growth, resource exploitation, and environmental perturbations. As Michael Colby among others has pointed out, this mental model has to some degree been internalized by much of modern mainstream neoclassical economics, which, by focusing on allocations of scarce resources and concomitantly treating environmental systems as infinite, treats environmental issues as essentially irrelevant. This position is sometimes referred to by ecological economists as "frontier economics" or "cowboy economics."

At the opposite extreme, there are those who believe that nature is extremely fragile, and that any change is likely to cascade out of control (Figure 3.2 b). This tends to be reflected in a radical ecology approach that supports an immediate return to low technology. The third ideology basically views nature as robust, but within distinct limits that, when transgressed, can result in a sudden shift to another metastable state (Figure 3.2 c). Depending on how stable such states are assumed to be, and the ideological predilections of the individual, this may result in positions ranging from "deep ecology" to industrial ecology (refer to Table 3.1). While both of these positions enable technological evolution, they differ significantly in their implications for the future structure of society. Deep ecology tends to envision a less complex world supported by "low tech" appropriate technologies, whereas industrial ecology envisions a more complex world in which technology, including information technology linking together previously disparate economic and environmental systems, is even more dominant than today. Thus, although similar in immediate implication, deep ecology and industrial ecology are fundamentally different.

These ideological positions derive from many sources, and each is supported by powerful belief systems and institutions within society. The danger for the student is not in the ideological nature of these positions themselves; personal and cultural values and ideology are critical dimensions of any dialog on sustainability. Rather, the danger arises when ideological conflicts are disguised as objective or factual disagreements. More data and research, and rational argument, can resolve the latter, but are of virtually no value in resolving the former, and the attempt to make them do so not only wastes resources, but can generate conflict in itself. Thus, an important goal of any industrial ecology or sustainability policy should be to separate the objective from the normative: the former to be researched, the latter to be addressed by the political and cultural processes characterizing that society.

3.4   THE MASTER EQUATION

Another way to illustrate the critical components of sustainability is by considering the "master equation," which links environmental impact with population, the quality of life sought, and the technology with which that quality of life is provided:

Environmental Impact = population × GDP/person × environmental impact/unit of GDP

where GDP is a country's gross domestic product, a measure of national economic activity. Let us examine each term more closely.

3.4.1   Environmental Impact

It is a common mistake to focus on any single regional or global environmental perturbation, such as stratospheric ozone depletion or global climate change, as exemplifying the environmental impact of the human species, and then, by arguing that the impact is either less than alleged or manageable, to imply that human environmental impacts are relatively trivial. Human environmental impacts, however, are profoundly systemic in nature, and cannot be simply controlled by mitigating a small set. To take a simple

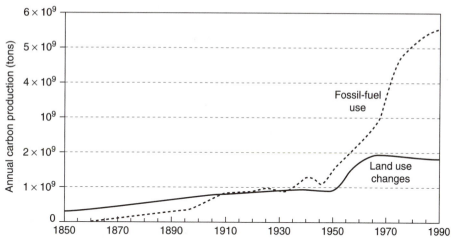

FIGURE 3.3   Annual Production of Carbon from Anthropogenic Fossil-fuel Use and Land Use Changes
Adapted from R.A. Houghton and G.M. Woodwell, "Global Climate Change," Scientific American, 260:4 (April 1989), 36–44.

example, the build-up of carbon dioxide in the atmosphere over the last several centuries is the result of two very different forms of human activity reflecting significantly different states of the global economy. Until about 1910, land use changes (primarily as a result of agricultural activities) were the dominant forcing factor; subsequently, however, fossil fuel use (a result of industrial and transportation technologies) predominates (see Figure 3.3).

Thus, the suite of regional and global impacts—including loss of biodiversity; disruption of habitat; increased global diffusion of toxics, both organic (e.g., persistent chlorinated compounds and pesticides) and inorganic (e.g., heavy metals and radioactive materials); loss of arable soil; loss and degradation of fresh water supplies; renewable and nonrenewable resource depletion; stratospheric ozone depletion; global climate change forcing; and increased acidity of precipitation—should be understood as indicia of one underlying force: the increased activity of a single species occurring over an extremely short time period (Figure 3.4). This dominance of global systems, appropriation of global resources, and domination of other species by a single species is unprecedented. Along these lines, Kates et al evaluated the rates of change in a number of critical anthropogenic (human-generated) environmental stressors, with 10,000 years before present being the zero point and 1985 being 100 percent, with the results shown in Figure 3.5: an extraordinary acceleration of impact within the recent past. This exponentially growing "global human forcing function," arising from the integration of human population and economic growth, is the disease; the familiar litany of environmental perturbations are only symptoms. They should, of course, be mitigated to the extent feasible; but this mitigation process should not be confused with addressing the underlying causal processes.

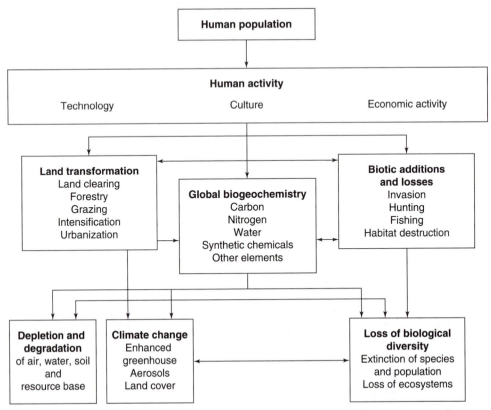

Human Impacts on Natural Systems: Conceptual Model

FIGURE 3.4 The important point is that human effects, however they are manifested, are a systemic impact. Human and natural systems are increasingly coupled together, especially at current and future scales of human activity.

Modified from P.M. Vitousek, H. A. Mooney, J. Lubchenco, and J.M. Melillo, "Human Domination of Earth's Ecosystems" Science 277: 292–499.

Tom Graedel has suggested a useful conceptual grouping of the suite of environmental impacts into four *Grand Objectives*. These are briefly presented in the four sections that follow; more detailed information on particular perturbations can be found in references provided at the end of the chapter.

3.4.1.1 Maintaining the existence of the human species.

Whether the anthropocentric bias of much policy is appropriate may be controversial, but its existence is not. For most people, this objective is and remains a primary concern. Among the environmental issues associated with this objective are global climate change; human health, safety, and mortality issues; water quality, quantity, and geographical distribution; resource depletion, especially fossil fuels; and distribution and management of radionuclides. Parenthetically, this category might be better thought of as maintaining a desirable, relatively stable condition for the human species; as

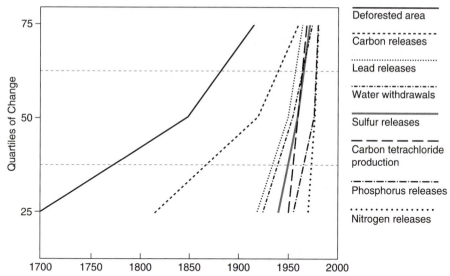

FIGURE 3.5   The scale of human impacts on natural systems is a relatively new phenomenon. Given the complex nature of those systems, it is likely that they may display unpredictable shifts as a result.
Adapted from R.W. Kates, B. L. Turner II, and W.C. Clark, "The Great Transformation," In The Earth as Transformed by Human Action (Cambridge Univ Press, Cambridge: 1990): 1–17.

noted earlier, it is unlikely the existence of the species is at risk, at least in the foreseeable future.

3.4.1.2   Developing and maintaining the capacity for sustainable development.

This involves two separate stages, subject to the fundamental problem of defining sustainable development. The first is developing the science and technology capability to define the concept; the role of policy here is essentially to support this R&D effort, and to develop the appropriate mechanisms for assuring that the results are integrated into the policy formulation and implementation process. The second, involving a heavier policy implementation dimension, is to develop and deploy policies to move toward and maintain a sustainable state. According to Graedel, among the environmental issues associated with this objective are water quality, quantity, and distribution; resource depletion, particularly of energy resources; and depletion of natural sinks at all scales, including landfills for solid wastes.

3.4.1.3   Biodiversity.

While there is some uncertainty about both the number of species existing now and the extent to which that number has been attenuated by human activity, there is less controversy over the general principle that loss of biodiversity—taken to mean both loss of individual species and loss of critical habitats and biological communities—is increasing. Contrary to popular belief, this is regarded as a serious issue by most people. For example, structured interviews carried out by Kempton et al. in the United States from a cultural anthropology perspective indicate fairly strong support for species preservation even among groups with incentives to oppose such a position (sawmill workers in the American Pacific Northwest):

> Survey question number 49 [with responses of target groups, percent agreeing with statement]: All species have a right to evolve without human interference. If extinction is going to happen, it should happen naturally, not through human actions.

Earth First!	Sierra Club	Public	Dry Cleaners	Sawmill Workers
100	82	87	77	59

Among the environmental issues that affect this objective are loss of biodiversity (species, habitat, critical communities); stratospheric ozone depletion; water quality, quantity, and geographical distribution; acid deposition; and thermal pollution.

3.4.1.4 Aesthetic richness.

Sustainable development and sustainability contain both objective and normative dimensions. Accordingly, it is appropriate to recognize that a world that is maintained in some sort of fashion, but which cannot provide quality of life, may be undesirable to most people, and that aesthetic considerations are a critical component of quality of life.

According to Graedel, concerns that impact aesthetic richness include, but are not limited to, smog; aesthetic degradation; oil spills; refuse disposal, formal and informal; odor pollution; and noise pollution.

3.4.1.5 Are current levels of environmental impact sustainable?

Given the lack of substantive data, it cannot be determined with certainty whether existing levels of human activity and associated environmental impacts are sustainable over the long term. Nonetheless, such data as are available, and the fact that many of these systems, human and natural, are quite complex and therefore quite likely to exhibit unpleasant discontinuities in their behavior, strongly indicate that, taken as a whole, the environmental impact of our species should, as insurance if nothing else, not just be slowed, but reduced in absolute terms. How to do that efficiently and effectively within the time and resource constraints that may exist (and most of which have yet to be identified) is yet to be resolved; indeed, the science and technology base necessary to phrase the questions properly, much less to answer them, has yet to be developed.

Limiting the environmental impact of human activity taken as a whole, however, will be no simple task, as the right side of the master equation indicates. The population and wealth per person terms are under strong upward pressure, with the latter perhaps even accelerating with the growth of the global economy. Whether and how the third term, which is basically a technology term, can compensate for this growth is the essence of industrial ecology.

3.4.2 Population

Human population growth is clearly an important contributing factor to increased anthropogenic environmental impact. It is generally recognized that human population growth has been explosive since the beginning of the Industrial Revolution, which, in conjunction with the concomitant revolutions in agricultural technology, international trade, forestry exploitation, and urbanization, in essence created unlimited resources for such a population explosion. What is less recognized is the degree to which globally sustainable levels of population appear to be correlated with technological and cultural evolution. Thus, Figure 3.6, based on work by Edward Deevey, Jr., shows that the three great jumps in human population have accompanied the initial development of tool use, the shift from hunter-gatherer cultures to agricultural cultures, and the Industrial Revolution. Such figures should perhaps be treated as illustrative rather than definitive, particularly as regards detail. Demographer Joel Cohen, in his recent authoritative work *How Many People Can the Earth Support*, for example, questions the adequacy of using models as an approach, declaring that (p. 96), "the available simple models do not describe human population history."

Two points can be made with a fair degree of confidence, however. The first is that, regardless of the specifics of the past history of human population, the Industrial Revolution essentially created unlimited resources for expansion of that population. The second is that, prior to the Industrial Revolution, population patterns tended to be independent: Different human populations grew or shrank in response to local and regional conditions. With the Industrial Revolution, however, population patterns from many different regions around the world—such as the Basin of Mexico, Egypt,

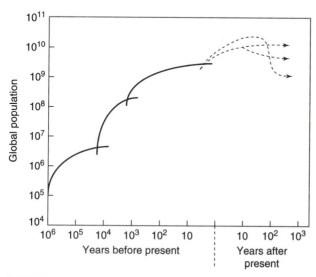

FIGURE 3.6 Population growth and stages of human evolution: the historical stages are tool use, the Agricultural Revolution, and the Industrial Revolution. The dashed lines show possible future paths: which is most likely cannot currently be predicted with any certainty.

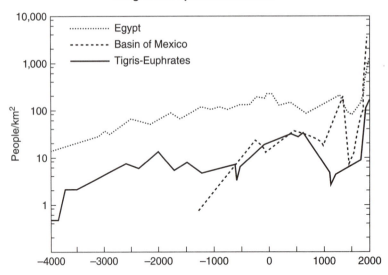

FIGURE 3.7 Note that the three populations evolve independently until around the beginning of the Industrial Revolution, then begin growing at unprecedented rates together.

Based on T.M. Whitmore, B.L. Turner II, D.L. Johnson, R.W. Kates and T.R. Gottschang, "Long Term Population Change," In The Earth as Transformed by Human Action (Cambridge Univ. Press, Cambridge:1990) pp.25–39.

and Mesopotamia—begin to move strongly upward in concert (Figure 3.7). What were disconnected systems became connected subsystems of the global population structure as a whole. As with environmental perturbations, this linked global human population system is a relatively new pattern—a new, and higher level, system in the world—and explains why previous experience may not be applicable in the future.

The human population continues to expand rapidly. Since 1970, human population has grown eight-fold. It is now approximately 5.7 billion, and projections of a peak of 8 to 12 billion or even more in the next century are not regarded by demographers as unrealistic. The timing and height of the eventual peak are difficult to predict with accuracy, and involve not only a number of cultural and political issues, but the fundamental, and unresolved, question of the earth's carrying capacity for our species given current technological systems and conditions.

What is clearly apparent, however, is that the first right-hand term in the master equation will be trending strongly upward unless some unforeseen catastrophe causes a population collapse. This is not only true because of the powerful institutions that remain pronatalist, but because of the demographic structure of the most rapidly growing populations, which tend to be in developing countries. Because these populations are usually heavily dominated by younger individuals just beginning to enter the reproductive cycle, population levels will continue to increase even if effective family planning techniques are fairly rapidly implemented. Moreover, to the extent that fundamental social changes—such as improvements in the status of women and economic development—turn out to be necessary to reduce population growth, it will obviously be some time before intentional population stabilization, if achievable at all, can occur.

3.4.3   Per Capita Wealth

Like the first term, the second term of the master equation is trending strongly upward. While wealth per capita varies significantly among countries and regions, and within political units may demonstrate substantial distributional disparities, the general trend is clearly positive, as Table 3.2 demonstrates. Citizens of developed countries are currently unwilling to significantly modify their consumption patterns for environmental reasons, nor are they willing to impose substantial financial penalties, such as taxes or fees, on environmentally inappropriate behavior such as inefficient energy use. Citizens of developing countries, exposed to consumption-intensive lifestyles through the global media, are unwilling to limit their aspirations for a similar lifestyle.

As with the parameters illustrated in Figure 3.5, it is seldom recognized how substantially economic activity has grown since the Industrial Revolution. Since 1700, the volume of goods traded internationally has increased some 800 times. In the last 100 years, the world's industrial production has increased more than 100-fold. In the early 1900s, production of synthetic organic chemicals was minimal; today, more than 225 billion pounds per year are produced in the United States alone. Moreover, many of the thousands of new materials introduced into commerce annually are not widely present in nature, so the eventual impacts of their use and dispersion, particularly on non-human species, may not be well characterized. Since 1900, the rate of global consumption of fossil fuel has increased by a factor of 50—much of it for transportation uses. What is important is not just the numbers themselves, but their magnitude and the relatively short historical time they represent. Moreover, in the past 15 years alone, economic activity has become far more global (Figure 3.8), creating, as in the cases of population and environmental perturbations, a system of global scale out of previously weakly linked separate regional economic systems.

At the present time, quality of life, which is what the second term really refers to, is almost universally defined in terms of ability to appropriate goods and materials. Owning things, especially iconic articles such as automobiles, is an important goal for

TABLE 3.2   Growth of Real per Capita Income in More Developed and Less Developed Countries, 1960–2000*

Country Group	1960–1970	1970–1980	1980–1990	1990–2000
More Developed Countries	4.1	2.4	2.4	2.1
Sub-Saharan Africa	0.6	0.9	-0.9	0.3
East Asia	3.6	4.6	6.3	5.7
Latin America	2.5	3.1	-0.5	2.2
Eastern Europe	5.2	5.4	0.9	1.6
Less Developed Countries	3.9	3.7	2.2	3.6

*The figures are average annual percentage changes; and for 'Less Developed Countries,' entries are weighted by population. The 1990–2000 figures are estimated.
Source: Data from The World Bank, World Development Report 1992. Oxford, UK: Oxford University Press, 1992.

Percent of World GDP and Population Linked by International Trade

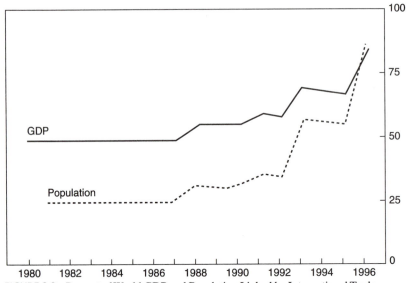

FIGURE 3.8 Percent of World GDP and Population Linked by International Trade
Source: Based on The Economist, June 14, 1997, p. 19.

TABLE 3.3 Extension of U.S. Commodity Consumption Patterns to the Global Economy

Commodity	1990 U.S. Apparent Consumption	1990 World Production	Necessary World Production for World per capita Consumption to equal U.S. per capita	Factor Increase
Plastic	25.0	78.3	530.0	6.8
Synthetic fibers	3.9	13.2	82.7	6.3
Aluminum	5.3	17.8	111.5	6.3
Copper	2.2	8.8	46.0	5.2
Salt	40.6	202.3	860.7	4.3
Potash (K_2O Content)	5.5	28.3	115.5	4.1
Industrial sand and gravel	24.8	133.1	525.3	4.0
Iron and steel	99.9	593.7	2,117.9	3.6
Nitrogen	18.0	107.9	381.0	3.5
Cement	81.3	1,251.1	1,723.1	1.4

Source: Based on U.S. Bureau of Mines data; calculations based on 1990 U.S. population of 249 million and world population of 5.3 billion.

most of humanity. As Table 3.3 indicates, however, it is unlikely that the globe, already under significant stress from current levels of material consumption, would be able to support a world economy based on today's production and consumption levels. This may not be for reasons of absolute scarcity: New sources of materials and substitutes for some uses will be found as prices shift. Rather, it is because the cumulative environmental impact of this amount of material cycling through the world economy—including associated activities such as transportation, packaging, energy consumption and maintenance—will destabilize or otherwise unacceptably perturb existing natural systems.

Quality of life, of course, is not necessarily the same as GDP. The link between material possessions and even survival is a clear one in a primitive economy, and the origins of cultural and social patterns that equate success with consumption and material possession are both apparent and rational. In a modern developed economy, however, consumption levels are far above any required for survival or even, in many cases, a comfortable existence. In fact, to the extent such patterns result in environmentally unsustainable economic behavior, they are profoundly unadaptive. This implies that at some point the link between perceived quality of life and material consumption must be weakened or broken. Interestingly enough, there is some indication that, in fact, happiness has not accelerated with material consumption in the United States after World War II (Figure 3.9), which seems to imply some degree of decoupling. If decoupling has, in fact, occurred, however, it has not reduced material consumption noticeably, at least in the short term. Thus, changing existing patterns of consumption must be regarded as ideologically and politically difficult, especially in the short term. For now, therefore, the trend of the second term must be accepted as being strongly upward.

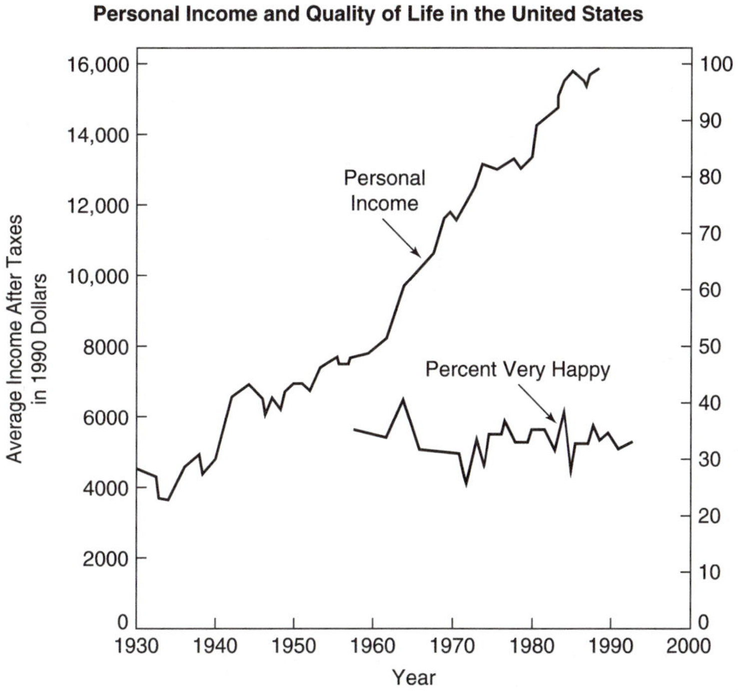

Personal Income and Quality of Life in the United States

FIGURE 3.9   While personal income and the quality of life in the United States have apparently been decoupling since the end of World War II, the effect has not been to reduce per capita consumption of materials.

Source: Based on D.G. Myers and E. Diener, "The Pursuit of Happiness," Scientific American, 274 (5): 70–72 (May 1996).

3.4.4   Environmental Impact per Unit of Production

The third term in the master equation is the only one that, at least in the short term, offers any possibility of reducing the fundamental environmental impacts of human economic activity. Defined as the environmental impact per unit of gross domestic product, it can be considered as the environmental impact necessary to provide a unit of quality of life to an individual. It represents the degree to which technology is available to meet human needs and aspirations without serious environmental consequences.

It is important to realize that the evolution of more environmentally appropriate technology is not a substitute for population stabilization and cultural change. Rather, technology—broadly defined as "the means by which a society provides quality of life to its members"—is the principle system that can be evolved in the relatively short term to mitigate environmental impacts, at least in some cases. An example is provided by the effort to reduce emissions of chlorofluorocarbons, which are implicated in the destruction of the stratospheric ozone layer. One use of such materials was in automobile air conditioning systems. Finding a technological substitute—in this case, a combination of other materials with far less ozone depleting potential—and introducing it in new automobiles took only a few years. Had the alternative been to eliminate automobile air conditioners—a culture change—it would have been far more difficult to achieve public acceptance of the need to reduce emissions of CFCs.

There are four critical dimensions of technology, taken in this broad sense, for sustainability:

1. Technology must always be thought of as part of a system, which includes not just the artifact(s) in which technology is physically embodied, but also the psychological, cultural, and economic systems within which that technology is embedded and used.

2. No technology should be considered good technology unless it is as environmentally preferable as possible within its economic and cultural context. The state-of-the-art does not yet permit the definition of technologies, artifacts, or technological systems as "sustainable," but enough is known already to make environmentally preferable choices in many instances, at least where complex or difficult trade-offs among impacts are not involved. The sum of such decisions will move the global economy in the right direction.

3. By defining technology as the way in which a society provides quality of life to its members, continued development can be conceptually separated from ever-increasing levels of consumption of raw materials and energy. Once a minimal level of well-being can be obtained, the focus of technological evolution shifts from increasing material wealth to providing more highly valued functionality, a higher quality of life.

4. Implicit in this approach to technology is the idea that it is not enough to simply develop a good technology. No technology will produce gains in economic and environmental efficiency until it is diffused throughout the economy, and is actually being used to provide quality of life to members of a society. In many cases, technology development is relatively simple compared to the difficulties inherent in technology diffusion. Perhaps reflecting this, there is a far more developed lit-

erature on technology development than technology diffusion, and no good theoretical basis for understanding technology diffusion has yet been developed.

The third term of the master equation reinforces the critical point that, unless one is implicitly or explicitly willing to accept unplanned reductions in human population, technology must be evolved that will continue to meet economic and cultural needs, yet do so in a more environmentally appropriate way than those that evolved during the Industrial Revolution (and reflect the lack of scarcities, and unconcern with any environmental impacts, that characterized that period).

It is important to note that this focus on technology should not be confused with naive technological optimism, which would hold that technological fixes will always mitigate any unforeseen environmental impact or continued uncontrolled growth of population. Rather, it is a necessary realism to accept the world as it is—with a growing human population desiring higher levels of material wealth—and try to evolve human systems to support a rational path toward long-term sustainability under these constraining conditions. In the immediate future, only rapid technological evolution appears to be able to fill that role. It will be quite difficult, and take considerably longer, to achieve the social and cultural evolution that will be required to stabilize population growth and change consumption patterns that link material wealth and quality of life. In fact, the data do not definitively support even the guardedly optimistic assumption that, by immediate implementation of industrial ecology, economic, cultural, or demographic disruptions arising from environmental perturbations can be avoided. Rather, in the spirit of William James's pragmatism, industrial ecology adopts as an operative assumption the possibility of a reasonably smooth transition to a stable carrying capacity. To assume otherwise runs the risk of either despair or extremism, and the creation of a self-fulfilling prophecy. It is not optimism, then, but realism that recognizes the critical role of industrial ecology.

3.4.5 Culture and Technology

The master equation thus illustrates in more detail the important conceptual separation between the scientific-technological and cultural-political dimensions of sustainable development illustrated in Figures 2.1 and 3.1. This leads to several principles regarding the interplay between culture and technology:

1. No technology or technological system exists in a vacuum; they are co-evolved with linked technologies and become embedded in social and cultural patterns. These co-evolved systems exert a strong influence on the speed with which any technology can be evolved.

2. Although there are exceptions, in general, cultural change is more difficult than technological change. In particular, product and process changes involving technologies that are only weakly linked to other technological systems, and do not require significant changes in human behavior, may be accomplished relatively easily.

3. New technologies can be relatively easily developed, a technological and scientific process, but are much harder to diffuse into the economy, a process which

has far more cultural content. Moreover, cultural barriers to different technologies are often both high and subtle, making them difficult to identify and, even when identified, difficult to overcome.

4. Technological evolution can, in the short term, buy time for cultural evolution, but it is not a substitute. Better gas mileage for sedans, linked with lower emissions per mile, have not dramatically reduced the environmental impacts of tailpipe emissions in developed countries, because people are driving their cars farther and choosing less fuel efficient models, such as four-wheel drive sports/utility vehicles, rather than sedans. Better technological performance of the artifact—the family car—has to a large degree been negated by changes in consumer behavior and demand, which are primarily a cultural phenomenon.

5. Any policy intended to reduce the environmental impacts of major technological systems must include planning for cultural as well as technological evolution. In particular, policies aimed at fostering sustainability should adopt a systems view that integrates both the cultural and technological dimensions of the system.

3.5 HUMAN CARRYING CAPACITY AND SOCIAL FREE WILL

3.5.1 Human Carrying Capacity

Another way to think of sustainability is as the carrying capacity of the earth for the human species. *Carrying capacity* in biological ecology is the population size of a species that can be supported in a given environment (the number of trout in a stream, rabbits on an island, or flour beetles in your home flour container, for example). It is not simple to define carrying capacities for non-human populations, especially as conditions change over time (flour beetles consume the flour, for example, and their population then collapses unless more flour is provided).

The concept as applied to humans, however, is even trickier, primarily because technology and culture have intervened between people and the natural systems that would otherwise control their numbers. Accordingly, the carrying capacity for our species is not even conceptually a single number or simple function; rather, it is probably what mathematicians call a complex surface. In other words, there are a number of worlds that might well be maintainable over some time frame—sustainable in some sense of the word—and it is up to us to choose among them.

In his book *How Many People Can the Earth Support?*, the demographer Joel Cohen raises a number of questions that illustrate the choices involved:

1. How many [people] at what average level of material well-being?
2. How many with what distribution of material well-being?

3. How many with what technology?
4. How many with what domestic and international political institutions?
5. How many with what domestic and international economic arrangements?
6. How many with what domestic and international demographic arrangements?
7. How many in what physical, chemical, and biological environments?
8. How many with what variability or stability?
9. How many with what risk or robustness?
10. How many for how long?
11. How many with what values, tastes and fashions?

The obvious implication of this is that humans as a group have options, choices about where they as a species want to reside on the "sustainability surface." The choice may be conscious: a result of understanding and applying the precepts of industrial ecology, for instance. Alternatively, the species can simply choose not to choose, achieving what will inevitably be a less biodiverse world where, at least in poorer areas of the globe, human population is controlled by mortality. However, it is worth noting that, in many cases, individuals (as opposed to institutions or the species as a whole) will not be able to choose along these dimensions. Rather, individuals will be locked into roles by their cultures, opportunities, and personal situations: A marginalized subsistence farmer in a poor developing country will certainly not have the flexibility that a middle class citizen of a developed country would. Moreover, an obvious but necessary caution is that the data and methodologies that would enable such choices to be rationally made do not, in fact, yet exist. Even where they do, political and cultural constraints may prevent a rational response. These issues approach a difficult concept, that of *social free will*.

3.5.2 Social Free Will

It is worth quoting at length Joel Cohen's vision of the decision-making tool that is required to respond to such a challenge (and one that he freely admits is currently far beyond the "present primitive state of human intellectual capacity and knowledge") (p.359):

> An ideal tool for estimating how many people the earth can support would be a model, simple enough to be intelligible, complicated enough to be potentially realistic and empirically tested enough to be credible [footnote omitted]. The model would require users to specify choices concerning technology, domestic and international economic arrangements (including recycling), domestic and international demographic arrangements, physical, chemical and biological environments, fashions, tastes, moral values, a desired typical level of material well-being and a distribution of well-being among individuals and areas. Users would specify how much they wanted each characteristic to vary as time passes and what risk they would tolerate that each characteristic might go out of the desired range of variability. Users would state how long they wanted their choices to remain in effect. They would specify the state of the world they wished to leave at the end of the specified period. The model would first check all these choices for internal consistency, detect any

contradictions and ask users to resolve them or to specify a balance among contradictory choices. The model would then attempt to reconcile the choices with the constraints imposed by food, water, energy, land, soil, space, diseases, waste disposal, nonfuel minerals, forests, biological diversity, biologically accessible nitrogen, phosphorus, climatic change and other natural constraints. The model would generate a complete set of possibilities, including human population sizes, consistent with the choices and the constraints.

Consider this model in light of the discussion of complex systems in Chapter 9, and particularly in light of the comments of Senge and Sterman about the growing disparity between the cognitive skills and capabilities of individuals, and the complexity of the economic, technological, cultural, and physical environment within which they are expected to act. Under these circumstances, one must ask whether the individual, or a higher level social organization, is the level at which such choices will—perhaps must—be made.

Sustainability thus raises the interesting issue of free will—the ability to choose among options without the outcome being predetermined by physical, cultural, psychological, or other constraints—on a societal, rather than individual level. This is a difficult concept: Free will usually presupposes a sense of self, which creates a self-consciousness that is able to act freely within certain boundaries (what those boundaries are is a subject that has fascinated philosophers for long years). Can institutions within a culture or society, or the culture or society itself, or even the species act in such a way that an external observer would perceive a sense of self at that level, and the "conscious" exercise of free will in choosing among alternative futures?

Some specific dimensions of an inquiry into social free will can be identified. (In doing so, it will appear that "society" is being somewhat personified, but the possible problems involved in doing so are outweighed by the greater ease in understanding some of these concepts which results.)

First, as is discussed in more detail in Chapter 8, there are inadequate data and methodologies to understand the scientific and technological questions, much less the necessary answers, underlying the concept of sustainability. Moreover, the intellectual scaffolding necessary to support development of this knowledge base is just being created: indeed, this is the field of industrial ecology. Free will of any sort presupposes the necessary knowledge; social free will, if it exists, will require such knowledge to act. The trick is that knowledge of that type may be sufficiently complex that it is beyond any single individual, and can only be "known" at an institutional or organizational level.

Second, as Niklas Luhmann has pointed out, social institutions, like people, have limited perception and bounded responses: They can only detect certain kinds of problems, and they can only react to them in certain ways. For example, the legal system detects violations of laws and regulations, and has a defined set of responses (e.g., a lawsuit or enforcement action); it also functions within the more subtle behavioral and psychological constraints inherent in its particular society. Law is a very different practice in France, in the United States, and in the former Soviet Union. The legal system will respond to an illegal dumping of heavy metals in drinking water; it will "not see" clear cutting a virgin forest if the activity is otherwise legal, regardless of the environmental impact. In a sense, this is why there is much effort by environmental groups to define certain behaviors as "unlawful"; it is an attempt to structure classes of situations deemed inappropriate into a form that the social institution of the law can both "see" and respond to.

The limits of social perception are matched by constraints on communication among social entities. Boundaries between disciplines are strong, enforced not just by culture and reward systems, but by development of arcane professional languages and vocabularies that clearly delineate intellectual tribes. Governments, research organizations, funding structures, policy systems: All develop their own culture, their own focus, and structures that encourage internal communication, but act as barriers to inter-group communication. No one who has worked in academia or watched negotiations between different interest groups—trade professionals and environmentalists, or environmentalists and national security professionals, for example—can fail to understand this dynamic. Moreover, the increasing complexity of social and economic systems means that communications barriers may well increase, rather than decrease, over time, at least as between entities (as opposed to within entities, where communications tools such as the Internet may well enhance communications).

There are thus at least four interacting factors limiting the capability of social institutions and society itself to respond, particularly in the short term: 1) lack of knowledge; 2) limitations on ability to perceive problems; 3) limitations on responses that can be made to perceived problems; and 4) limitations on communication. A number of questions arise logically from this brief assessment:

- *How rapidly can society understand the situation?* In this case, it is apparent that the complex systems—physical, cultural, and economic—that are the subject matter of industrial ecology, and lie at the base of sustainability, are not well understood at all, but, dangerously, many people think they are.

- *How rapidly can society understand the options before it?* This question goes to the issue of the capability of social institutions to perceive the real dimensions of these issues, and communicate in such a way as to understand what is perceived.

- *How free is society to respond to this understanding, even if it can be achieved?* For example, consider the strong position against family planning and equal rights for women taken by some major religious groups, and how difficult those would be to change even if society recognized in general the real costs and dangers of continued human population growth. In the economic sphere, consider the opposition to global climate change science and policy by national states and sectors that produce fossil fuels. Such opposition need not be unprincipled; it may reflect interpretations of data and political factors that are strongly influenced by cultural and psychological models arising from the role of the individual and organizations within society. Regardless, however, they constitute real, and generally poorly understood, barriers to social responses.

- The preceding question leads logically to another question: *How rapidly can society evolve to respond (and can the evolution of social responses be synchronized with the evolution of the natural systems under human forcing, and if so, how)?*

  More specifically, are there identifiable organizations within society (transnationals, the national state, environmental non-governmental organizations [NGOs]) that have the authority and capacity to identify options and act upon them appropriately? This question raises both practical and theoretical difficulties. The practical difficulty is that, for each institution that can be identified, a significant constraint upon its ability to understand, perceive, and act can easily be found. NGOs are frequently captured by their cultural models; national states, an obvious choice, frequently have their governmental apparatus captured by elites and special interests; and transnationals are limited by the very legal structure that creates them (see Chapter 16).

  The theoretical difficulty arises from the fact that sustainability, if it is achieved, will be expressed at the level of human society as a whole, not at subsystems level. Therefore, just as an individual might not be able to exercise choice at the level of global sustainability options (might have no free will in the matter), neither might an organization or institution that, like an individual, is also only a subsystem of the larger, global system.

- A final question, which raises a number of interesting issues in its own right, is the extent to which individuals could even consciously perceive, much less understand, free will existing at a social level and, if they could, whether or how they could affect its exercise.

These are deep waters, and it is as well to bring the reader back to shore before the chapter ends. It is important to understand that the carrying capacity of the world for the human species is not predetermined, nor is there a single optimum; rather, it involves a choice, intentional or unintentional, among many possible options. Whether, and how, these choices can be made in practice is less clear, and involves complicated questions of social institutional capacity, institutional evolution, and complex systems dynamics.

REFERENCES

Brown, L. R. and the staff of the Worldwatch Institute. *State of the World.* New York, W. W. Norton and Company: 1996.

Choucri, N., guest editor. *Business & The Contemporary World.* (special edition on Global Environmental Accords: Implications for Technology, Industry and International Relations), VI(2), 1994.

Cohen, J. E. *How Many People Can the Earth Support?* New York, W. W. Norton & Company: 1995.

Colby, M. E. "Environmental Management in Development: The Evolution of Paradigms," World Bank Discussion Paper No. 80, 1990.

Daly, H. E. and Cobb, J. B. Jr. *For The Common Good: Redirecting the Economy Toward Community, the Environment, and a Sustainable Future.* Boston, Beacon Press: 1989.

Luhmann, Niklas. *Ecological Communication.* Translation by J. Bednarz, Jr., published by University of Chicago Press: 1989; original published in the German by Westdeutsches Verlag GmbH, Opladen: 1986.

Navarro-Valls, *Wall Street Journal.* p. 12, September 1, 1994.

Phipps, E. "Overview of Environmental Problems," *National Pollution Prevention Center for Higher Education.* University of Michigan, Ann Arbor, MI: 1996.

Robey, B., S. O. Rutstein, and L. Morris. "The fertility decline in developing countries." *Scientific American.* 269(6): 60–67, December 1993.

Turner, B. L. II, W. C. Clark, R. W. Kates, J. F. Richards, J. T. Mathews, and W. B. Meyer, eds. *The Earth as Transformed by Human Action.* Cambridge, England, Cambridge University Press: 1990.

Whitmore, T. M., B. L. Turner II, D. L. Johnson, R. W. Kates, and T. R. Gottschang, "Long-term population change." in Turner, B. L. II, W. C. Clark, R. W. Kates, J. F. Richards, J. T. Mathews, and W. B. Meyer, eds. *The Earth as Transformed by Human Action.* Cambridge, England, Cambridge University Press: 1990.

Weiss, E. B. *In Fairness to Future Generations.* Dobbs Ferry, NY, Transnational Publishers Inc.: 1988.

World Commission on Environment and Development. *Our Common Future.* Oxford, Oxford University Press: 1987.

EXERCISES

1. You are newly appointed the Research Director for Sustainable Development for the Netherlands, and given a budget of 40 million dollars per year. Develop a research agenda for your operation, including the strategic goals of your overall program, the program areas in which you will invest to meet those goals, and the projects that you will fund under each program. Also identify (by function) the critical ministries with which you must collaborate (for example, the Environmental Ministry, the Transportation Ministry). Allocate your resources among these projects.

2. Perform the activities in exercise 1, but assume your country is:

   a. Costa Rica

   b. The Peoples Republic of China

   c. The United States

   How do your answers for each country differ? Do you think that sustainability means different things depending on a country's state of development? If so, can global sustainability be achieved simply by summing up the efforts of individual countries? Why or why not?

3. Pick a room of your house, apartment, or dormitory. Inventory the physical items in the room. Divide them into four categories: 1) artifact is necessary for survival; 2) function is necessary, but artifact represents unnecessary environmental impact (e.g., clothes may be necessary, but a fur coat or 10 pairs of shoes may not be); 3) artifact is unnecessary for survival but culturally required; and 4) artifact is both physically and culturally unnecessary (albeit presumably desirable, or it wouldn't be there). Can you extrapolate this result to your general consumption patterns? Based on these results, what percent of your consumption represents unnecessary environmental impact?

4. Describe a sustainable world in your own words. Include a description of the lifestyle you would expect in such a world, as well as estimates of how large a population could be supported in such a world. What data and analysis would you need to be confident that your vision was, in fact, sustainable?

CHAPTER 4

Industrial Ecology

4.1 DEFINITION OF INDUSTRIAL ECOLOGY

The essence of industrial ecology, as defined by Graedel and Allenby in the first textbook in the field, may be simply stated as follows:

> Industrial ecology is the means by which humanity can deliberately and rationally approach and maintain a desirable carrying capacity, given continued economic, cultural, and technological evolution. The concept requires that an industrial system be viewed not in isolation from its surrounding systems, but in concert with them. It is a systems view in which one seeks to optimize the total materials cycle from virgin material, to finished material, to component, to product, to obsolete product, and to ultimate disposal. Factors to be optimized include resources, energy, and capital.

The words "deliberately" and "rationally" indicate that the intent of the multidisciplinary field of industrial ecology is to provide the technological and scientific basis for a considered path toward global sustainability, in contrast to unplanned, precipitous, and potentially quite costly and disastrous alternatives. "Desirable" indicates that, given the potential for different technologies, cultures, and forms of economic organization, a number of sustainable states may exist. It then becomes a human responsibility to choose among them, and then act so as to approach a desired state that otherwise would not have occurred.

The electronics industry, especially in the United States, has been a leader in developing industrial ecology and exploring its implementation through *Design for Environment (DFE)* practices. Accordingly, it is instructive to quote the definition provided in the "White Paper on Sustainable Development and Industrial Ecology" issued by the Institute of Electrical and Electronic Engineers, Inc. (IEEE):

> *Industrial ecology* is the objective, multidisciplinary study of industrial and economic systems and their linkages with fundamental natural systems. It incorporates, among other things, research involving energy supply and use, new materials, new technologies and technological systems, basic sciences, economics, law, management, and social sciences. Although still in the development stage, it provides the theoretical scientific basis upon which understanding, and reasoned improvement, of current practices can be based. Oversimplifying somewhat, it can be thought of as 'the science of sustainability.' It is important to emphasize that industrial ecology is an objective field of study based on existing scientific and technological disciplines, not a form of industrial policy or planning system.

This elucidation makes the important point that industrial ecology strives to be objective, not normative. Thus, where cultural, political, or psychological issues arise in an industrial ecology study, they are evaluated as objective dimensions of the problem. For example, in studying the automotive technology system, as Graedel and Allenby do in *Industrial Ecology and the Automobile,* the fact that improvements to the environmental performance of the automobile as an artifact have been substantial, but are outweighed by changes in customer use and demand patterns (i.e., a shift in fleet composition toward less fuel-efficient four-wheel drive sport-utility vehicles and trucks, with each vehicle being driven farther on average than before) is noted and its implications explored. Whether this is good or bad—whether it "should" be the case—is not properly an issue for industrial ecology.

A full consideration of industrial ecology would include the entire scope of economic activity, such as mining, agriculture, forestry, manufacturing, and consumer behavior. Both demand side (consumer) and supply side (producer) aspects of economic behavior, and consequent impacts on natural systems at all temporal and spatial scales, are included. Equally as important, industrial ecology not only includes economic activity in advanced economies, but also subsistence human activity at the fringes of formal economic systems. After all, many critical impacts, both in terms of human health and ecosystem and biodiversity maintenance, are generated by activities at that level.

4.2 HISTORY OF INDUSTRIAL ECOLOGY

The term "industrial ecology" itself seems to have been first used in the title of a very limited circulation journal in 1970, of which only one issue apparently was ever printed. Judging by the subject matter, it was intended simply to reflect the fact that industrial activities had an impact on nature. In 1972, the Ministry of Industry and International Trade in Japan (MITI) began considering such a metaphor to suggest a model for structuring the Japanese industrial system; relevant documents are all in Japanese, and the effort was dropped with the coming of the energy crisis of 1973. In 1989, an article by Frosch and Gallapoulos revived the term as an analogy, suggesting that industrial systems could be more efficient if their material flows were modeled after natural ecosystems.

At the same time, several important studies of the interrelationship between technology and industrial activities and environmental impacts were being published. The U.S. National Academy of Engineering (NAE) was particularly active in this area, with noteworthy early efforts including *Energy and Environment* (1989) and *Energy: Production, Consumption, and Consequences* (1990). Another important contribution was *The Earth as Transformed by Human Action* (1990), a comprehensive and still valuable source. Although these earlier publications did not use the term "industrial ecology," unlike much of the literature published in the environmental sciences, they did not take the reductionist approach, but attempted to view the activities from a more comprehensive, systematic basis—that is, they began to take a true industrial ecology approach to the issues.

Meanwhile, the complex, systems-based nature of regional and global environmental perturbations—particularly global climate change, ozone depletion, and loss of biodiversity—and the growing inability of existing environmental policies based on reductionist approaches to address such issues adequately became increasingly obvious. Additionally, industry was increasingly affected by rapidly rising costs of environmental compliance and cleanup, which were perceived as highly inefficient both environmentally and economically. These linked pressures generated a strong need for a new paradigm, a new way of thinking about these issues. Concomitantly, the 1987 report from the World Commission on Environment and Development, *Our Common Future,* had begun the dialog on the concept of sustainable development, a goal that clearly pushed beyond the existing environmental regulation approach.

Responding in large part to these pressures, industry, and particularly the electronics industry in the United States, began to develop a set of practices based on industrial ecology principles. These were generally captured under the rubric of *Design for Environment,* to indicate that they were intended to be a module of existing concurrent engineering practices in that industry, which are based on a "Design for X" approach, where X is any desirable characteristic of the product being designed (manufacturability, safety, testability, and so on). Thus, in 1993, the American Electronics Association published a collected set of White Papers under the title *The Hows and Whys of Design for the Environment: A Primer for Members of the American Electronics Association.* More broadly, the Office of Technology Assessment of the U.S. Congress issued *Green Products by Design* (1992), which reinforced the focus on products, rather than specific materials or localized impacts.

Despite the OTA report, the product-oriented focus has been generally ignored in the U.S. It has, however, become a central tenant of European technology and environment policies, particularly in such Northern European countries as the Netherlands, Germany, and Sweden. Examples include the work on "Integrated Substance Chain Management" in the Netherlands (1991), and the report "Extended Producer Responsibility as a Strategy to Promote Cleaner Products" from Lund University in Sweden (1992). Perhaps the country that has advanced farthest in this area is the Netherlands, which is attempting to develop an integrated policy structure based on sustainability (see the case study in Chapter 17).

Simultaneously, the electronics industry was establishing unique organizations, such as the Industry Cooperative for Ozone Layer Protection, or ICOLP, to address specific issues (in this case, ozone depletion) in a much more proactive, non-compliance oriented way. Significantly, treatment of these issues was characterized by a focus on manufacturing technologies, products, and processes, rather than on environmental details, indicating that the process of transition of environmental concerns from overhead to strategic had begun. Substantial research efforts aimed at understanding the impacts of chemicals in particular applications over their life cycle began under the rubric of "life-cycle assessment"; the Society of Environmental Toxicology and Chemistry (SETAC) and a number of European organizations have been particularly instrumental in this development.

Even though the term "industrial ecology" was not initially used in Europe, much related activity occurred there, particularly in the Netherlands, Germany, and the Scandinavian countries. The Netherlands, for example, initiated a unique effort to define a sustainable society in one generation, publishing their first *National Environmental Policy Plan* in 1989, and the *National Environmental Policy Plan Plus* in 1990. European private industry also contributed significantly to the dialog on development and the environment; of particular note is *Changing Course,* by Stephan Schmidheiny and the Business Council for Sustainable Development, published in 1992.

The effect of these mutually reinforcing trends has been to evolve industrial ecology from just an interesting metaphor into a nascent field of study. While this process has not yet been completed, and the term is still subject to misuse, a number of new publications have begun to formalize the field. Among the more notable are *The Greening of Industrial Ecosystems* and *Industrial Ecology: U.S.-Japan Perspectives,* issued in 1994 by the NAE; *Industrial Ecology and Global Change,* an edited volume of papers from a 1992 workshop in Snowmass, Colorado, published in 1994; *Industrial Ecology,* the first engineering textbook in the field, published in 1995; and *Industrial Ecology and the Automobile,* an engineering case study textbook published in 1997. Importantly for the development of the field, a new scientific journal, the *Journal of Industrial Ecology,* based in the Yale University School of Forestry and Environmental Studies, began publication in 1997.

In parallel, and much in the spirit of the Frosch and Gallapoulos, there have been several efforts to begin to apply the models and understanding of biological ecology to industrial ecology questions, beginning with the article appropriately entitled "Understanding Industrial Ecology from a Biological Systems Perspective" by Allenby and Cooper in 1994, an article by Graedel "On the Concept of Industrial Ecology" in 1996, and the volume *Engineering Within Ecological Constraints,* published by the NAE in 1996. Such efforts, however, must always be undertaken cautiously and with care; the economics literature is replete with examples of biological analogy run amok ("The bond market is much like the lymphatic system . . . "). The point is not that economic systems are biological systems. They are not. The point is that much work on complex systems has been done in the context of biological ecology, and some of this learning is applicable to economic and industrial systems despite their differences—in fact, a principle value of the analogy may be to highlight some of those differences.

Reflecting its origins, much of the early industrial ecology literature tends to focus on manufacturing sectors, and the confusion about the term as merely an interesting analogy, as opposed to a legitimate field of study, has not all been dissipated. Nonetheless, a substantial body of scholarship on the fundamental focus of the field—the links between economic and industrial activity and underlying natural systems—has begun to develop, a process that will be significantly accelerated by the publication of the *Journal of Industrial Ecology.* The basic rationale for the field, captured in the definition discussed in the section above, also becomes stronger and more obvious as the cumulative effect of human activity on the environment continues to become more onerous and apparent.

4.3 INDUSTRIAL ECOLOGY MODEL SYSTEMS

The concept of industrial ecology can be illustrated by considering three different models of systems, based here on the biological analogy. A Type I system is linear: Virgin materials enter the system, are used only once, and are then disposed of as waste (Figure 4.1a). ("Waste" in this book refers specifically to material for which there is no further use within the system, as opposed to "residuals," which may have no further use within the generating process or firm, but could be reused within the system as a whole.) Thus, an example of a Type I material flow in an economy would be an automobile that is abandoned by the side of the road at the end of its useful life—a de facto waste at that point, although from a material perspective it should be a residual (i.e., the technology exists to recycle much of the hulk). It is a principle of industrial ecology that no economic activity should create waste, only residuals.

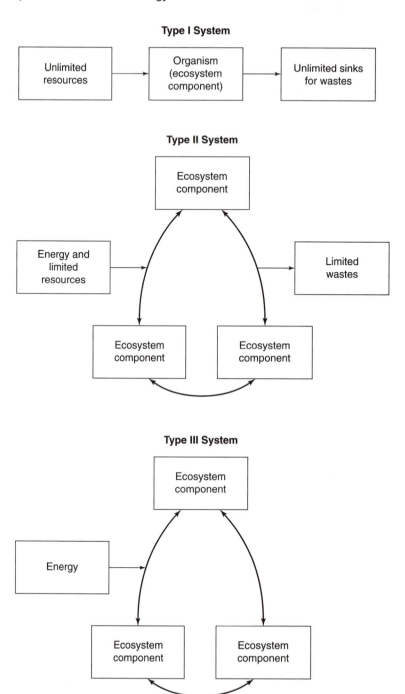

FIGURE 4.1 As one moves from a Type I to a Type III system, materials that were previously wasted become part of the system, which can then support more biomass or, in the case of the human economy, people.

A more complex Type II system arises as scarcity (or growing populations) makes the Type I system inadequate (Figure 4.1b). Feedback and internal cycling loops develop in large part through the evolution of new organisms (or products, subsystems, and activities within an economy). Accordingly, flows of material into, and waste out of, the system diminish on a per unit basis (e.g., per weight of biomass supported, or per dollar of economic activity). Internal reuse of materials can become quite significant. In an economy, while the internal material use within the system may remain high, the velocity of materials through the system is reduced. Material management systems, either planned or spontaneous, become more prevalent.

An example of a Type II system is the life cycle of most cars today in developed economies. The post-consumer car is first stripped of useful subassemblies, which are then reconditioned, if necessary, and recycled as used parts. The remaining hulk is shredded, and the steel, which is some 75 percent by weight of the automobile, is recovered for recycling. This is an internal material loop. The remaining plastic, glass, and miscellaneous materials and liquids, known as "fluff" or "ASR" (automobile shredder residue), are then landfilled. Landfills, which would be both environmentally and economically more efficient if conceptualized and designed as residual storage facilities as opposed to heterogeneous waste disposal sites, currently are not part of material cycling streams (except for a minor amount of methane from microbial degradation of carbonaceous material). Accordingly, the fluff stream constitutes waste: It would be more efficient if it were redesigned or managed so as to become a residual.

The Type III system is one in which full cyclicity has been achieved (Figure 4.1c). Such a system should, for many material streams and products, be the goal, but recycling is not necessarily always desirable, especially given current economic and technological conditions. For example, local energy recovery from plastics might well be preferable to shipping lightweight plastics long distances to reformulating facilities because of the environmental impacts of the transportation required in the latter case. Materials recycling must also be considered broadly, in conjunction with related technologies and natural systems. Thus, for example, production of energy from biomass feedstocks must be evaluated in the context of the global carbon cycle and the environmental implications of all sectors involved in the process (e.g., agriculture or forestry; agricultural chemical production, distribution, and application; biomass transportation and processing; energy production and transmission; and energy end-use efficiency).

Reverting to the automobile example, such a self-contained system should be the goal of automobile design and management across the life cycle, and, eventually, the goal of supporting infrastructure design as well. There are two important caveats to this blanket statement: The first is that recycling should be done only where it is environmentally efficient (that is, where the environmental impacts of recycling are less than alternatives, taken as a whole). In many cases, the data necessary to make this determination do not yet exist. The second caveat is that it must be economically viable under existing conditions (of course, these can be changed through adjustments to tax structures, fee systems, and the like).

Moreover, achieving a total Type III system for many technologies, including automobiles, will be extremely difficult because of *dissipative* uses of materials; that is, materials are degraded, physically dispersed, and lost to the economic system in the course of a single normal use. For example, tires leave dissipative residue on pavement, and fuel is burned in combustion engines. In such instances, minimization of dissipative uses, especially any dissipative uses of toxics or materials that may be toxics, may be the best that can be hoped for given current technologies. Use of lead in gasoline, for example, is obviously contraindicated.

In all such cases, it is important to recognize that there is no inherent requirement for the materials management function in any of these recycling systems to include only one firm or one sector, although in some cases such "tighter" loops may be more environmentally efficient. In principle, rather, the goal should be a self-contained global economy (Figure 4.2), with materials efficiency optimized over the larger system. Thus, for example, the concept of geographically co-located industrial material cycling systems, such as eco-parks, is interesting, but is only one possibility among many. To focus too intently on any one potential subsystem runs the risk of overlooking the need to improve system performance as a whole.

Notice also that all systems are energetically open, as solar energy is a constant input. Clearly, the constraint here is that energy use by the system as a whole must be sustainable; that is, the system must not rely on more energy than is constantly replenished by the sun (and, perhaps, nuclear fission and fusion should those technologies be developed to be environmentally and politically acceptable).

Simple as this discussion is, it illustrates several critical aspects of industrial ecology, which are covered in the following sections.

4.3.1   Systems Orientation

Industrial ecology is profoundly systems-oriented. This does not mean that every industrial ecology activity should have to include all possible impacts on any relevant

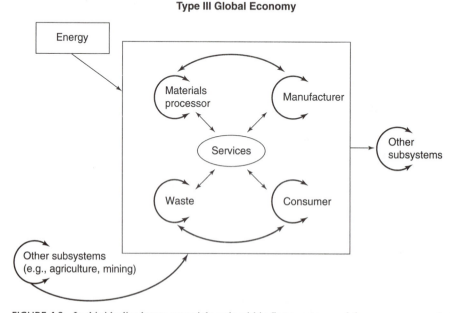

Type III Global Economy

FIGURE 4.2   In this idealized state, materials cycle within firms, sectors, and the economy as a whole. It is thus closed with regard to materials, but, like the earth itself, open energetically.

system; that would obviously be impracticable. It does, however, require a sensitivity to the systems aspect of industrial ecology research. Where boundaries are drawn around particular activities or subsystems, as they must be, they should be justifiable and reasonable in the context of the relevant system, and the interfaces between the subsystem and the external environment must reflect the integrated nature of the system as a whole. Of particular concern is the need to assure that "emergent behavior"—that is, behavior at a certain level of the system that cannot be predicted simply by summing up the activities at lower levels of the system—not be inadvertently overlooked.

Thus, for example, research on a carbon dioxide pellet cleaning system for metal pieceparts as part of a Design for Environment initiative should not have to include explicit consideration of the global climate cycle. It should, however, include sufficient research to demonstrate that, taken as a process step, it has a high probability of being environmentally preferable to the alternatives (as well as technologically and economically acceptable as well).

4.3.2 Complexity and Carrying Capacity

Increasing complex systems tend to be able to support increasing biomass, all else equal. This not only appears to be true for natural systems, as the work of Tillman et al. with regard to the positive relationship between biological productivity and diversity in grasslands suggests, but also to be true for human systems, as Figure 3.5 suggests. A hunter-gatherer society is more simple than an urban, industrialized society, but it can support far few people than the latter. This, of course, leads to the interesting hypothesis that appropriate deliberate increases in the complexity of the existing human economy, perhaps through increased information density, can increase the human carrying capacity of the globe, all else equal.

4.3.3 Scale Issues

It is critical to recognize that all cycles within the system tend to function on widely differing temporal and spatial scales, a behavior that greatly complicates analysis and understanding of the system. This can be particularly difficult because many important natural systems and cycles, such as the carbon cycle, operate on time scales of decades to centuries, a length of time that is considerably beyond the psychological time horizon of both people and societies (Figure 4.3). Much of the political and social concern involving nuclear power, for example, arises from the fact that some residuals are radioactive for centuries—and thus must be managed for centuries—and social and technological systems are not routinely designed to incorporate such long-term activity.

4.3.4 Co-evolution of Human and Natural Systems

Natural and human systems are not separate, they are intimately co-evolved and have been for centuries (Figure 4.4). Thus, for example, early human metallurgical activity in Rome and China perturbed global natural cycles of metals such as copper (Figure 4.5) and lead (Figure 4.6). Today, no corner of the earth's chemical or biological systems is free of anthropogenic components such as metals or persistent organic compounds. It is equally true that no human environment of more than trivial size is so sterile that one cannot wipe a surface and find microbes.

Human Psychology and Natural System Scale

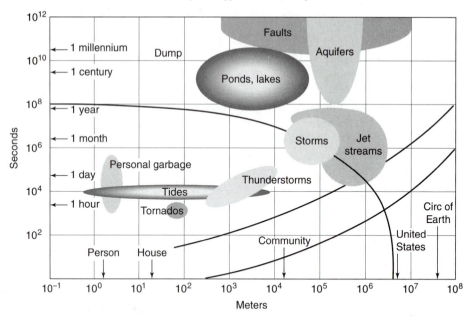

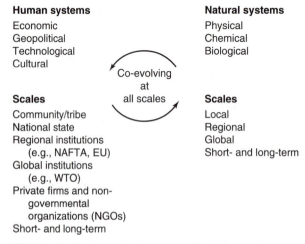

FIGURE 4.3   Many important natural cycles and systems lie beyond the usual human psychological horizon, and are thus both difficult to perceive, and to respond to.

Co-Evolution of Human and Natural Systems

Human systems	Natural systems
Economic	Physical
Geopolitical	Chemical
Technological	Biological
Cultural	

Co-evolving at all scales

Scales	Scales
Community/tribe	Local
National state	Regional
Regional institutions	Global
(e.g., NAFTA, EU)	Short- and long-term
Global institutions	
(e.g., WTO)	
Private firms and non-	
governmental	
organizations (NGOs)	
Short- and long-term	

FIGURE 4.4   Despite their intimate coupling, human and natural systems are usually perceived as separate.

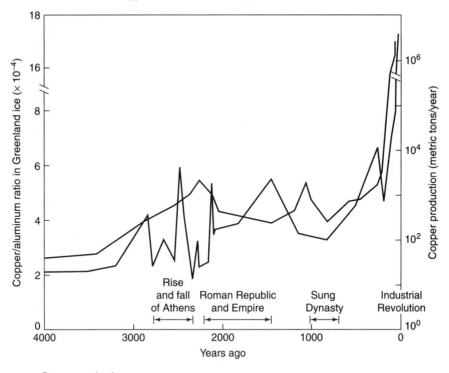

FIGURE 4.5   Note the similarity of this figure to those of regional populations in Figure 3.7: fluctuations until the Industrial Revolution, then rapid, unprecedented growth.

    Based on S. Hong, J. Candelone, C.C. Patterson, and C.F. Boutron, "History of Ancient Copper Smelting Pollution During Roman and Medieval Times Recorded in Greenland Ice," *Science* 272:246–249 (12 April 1996).

4.3.5 An Integrative Science

As the IEEE definition cited above makes clear, industrial ecology is an integrative, not a reductionist, field. It focuses on a comprehensive, holistic understanding of systems rather than the reductionist approach of developing more and more knowledge about increasingly specific subsystems, and, as Figure 4.7 illustrates, cuts across a number of disciplines in doing so. Figure 4.7 also illustrates another important characteristic of industrial ecology: the focus not just on individual systems, but on the links that exist among them (L1 through L8 in Figure 4.7). These may have a heavy cultural content as in the case of L7, which links cultural systems and the sustainability policy structure, or a heavy technology content as in L4, which links technological systems with industrial ecology, or L3, which links natural systems with industrial ecology. In all cases, however, the links exist between complex systems, incorporate a number of disciplines and dimensions, and cannot be wholly understood by taking the reductionist approach.

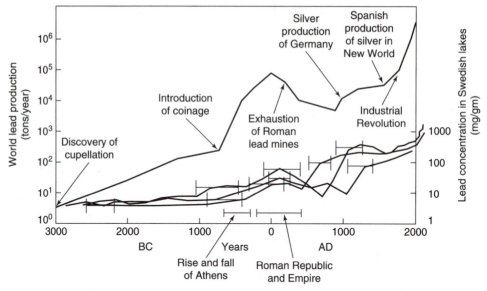

FIGURE 4.6 World lead production, and deposition in Swedish lakes. The close connection demonstrates that human economic activity, and global and regional environmental systems, have been coupled for many centuries. The Industrial Revolution, however, marks a watershed change in magnitude of impact.
Based on : Renberg, I, Persson, M.W. and Emteryd, O. Pre-industrial atmospheric lead contamination in Swedish lake sediments, Nature 363: 323–326.

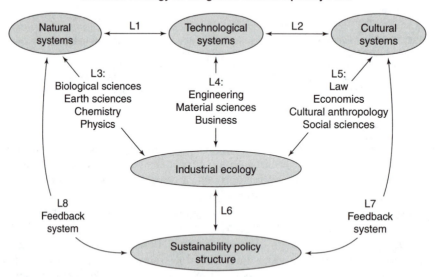

FIGURE 4.7 L1 through L8 mark important links between industrial ecology, and human and natural systems.

The Western scientific paradigm is, of course, highly reductionist, so it is not surprising that this facet of industrial ecology has a number of implications. For example, some scientists tend to view any approach that is not reductionist as lacking in scientific rigor or quality. More practically, the Western scientific edifice, from funding institutions to peer review systems, is not structured to support integrative disciplines. This practical consideration poses a substantial and very real barrier to the development of such fields. Analogously, virtually all colleges and universities are organized fairly rigidly around disciplines, making it difficult or impossible for practitioners of multidisciplinary fields, especially younger, non-tenured academics, to be supported. Obtaining tenure as a multidisciplinary scholar in a discipline-oriented academic institution, for example, is virtually impossible. It is important to recognize that disciplinary boundaries are extremely powerful: The strongest loyalty of academic communities is not to institutions, but to disciplines, and this is a global pattern (Table 4.1). These barriers are especially strong between social science disciplines, which study human systems, and the natural sciences (with engineering somewhat in the middle), which makes collaboration in the field of industrial ecology even more difficult (Figure 4.8).

4.4 PRINCIPLES OF INDUSTRIAL ECOLOGY

Although industrial ecology is a nascent field, a number of principles have already been developed. The most significant and all-encompassing of these, phrased to be applicable to a complex manufactured article, are presented below; analogous principles can be developed for other economic activities such as agriculture, forestry, or the production of personal-care products.

1. Products, processes, services, and operations can produce residuals, but not waste.
2. Every process, product, facility, constructed infrastructure, and technological system should be planned to the extent possible to be easily adapted to foreseeable, environmentally preferable innovations. For example, buildings should be designed to be capable of supporting photovoltaic systems by leaving an unshaded south-facing rooftop, even if such a system is not installed at initial construction.

TABLE 4.1   Academics' Loyalties (in percent[1])

Country	To Institution	To Discipline
Australia	74	94
Chile	95	100
Germany	34	91
Japan	80	97
South Korea	88	99
Russia	90	96
Sweden	56	89
United States	82	98

[1]Percent replying affiliation "very" or "fairly" important to them, versus "not very" or "not at all."

Source: data from *The Economist*, October 4, 1977 center section, "Universities", from survey by Carnegie Foundation

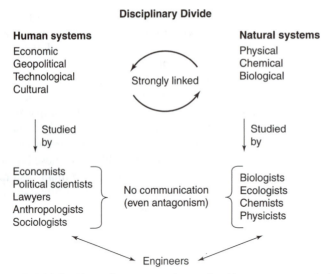

Disciplinary Divide

Human systems
Economic
Geopolitical
Technological
Cultural

Strongly linked

Natural systems
Physical
Chemical
Biological

Studied
by

Studied
by

Economists
Political scientists
Lawyers
Anthropologists
Sociologists

No communication
(even antagonism)

Biologists
Ecologists
Chemists
Physicists

Engineers

FIGURE 4.8 The tendency to regard natural and human systems as independent is institutionalized in the disciplinary structure by which they are studied.

3. Every molecule that enters a specific manufacturing process should leave that process as part of a salable product. This means, for example, that manufacturing processes should be designed not just to produce the primary product efficiently, but also residual streams which can be reused or sold.

4. Every erg of energy used in manufacture should produce a desired material transformation.

5. Industries should make minimum use of materials and energy in products, processes, services, and operations.

6. Materials used should be the least toxic for the purpose, all else equal. Unfortunately, the cases where all else is equal are usually trivial; it is far more common to have difficult trade-offs. For example, in many cases involving manufacture of complex objects, experience indicates that there is a trade-off between energy consumption and toxicity of process materials. In such cases, an abbreviated life-cycle analysis might be appropriate to determine whether it is at least likely that substitution of a more energy-intensive, but less toxic, material will indeed be environmentally preferable.

7. Industries should get most of the needed materials through recycling streams (theirs or those of others) rather than through raw materials extraction, even in the case of common materials.

8. Every process and product should be designed to preserve the embedded utility of the materials used. This might involve designs which extend the life of the product, or facilitate recycling of subassemblies or components, rather than just materials. An efficient way to accomplish this goal is by designing modular equipment and by remanufacturing.

9. Every product should be designed so that it can be used to create other useful products at the end of its current life.

10. Every industrial landholding, facility, or infrastructure system or component should be developed, constructed, or modified with attention to maintaining or improving local habitats and species diversity, and to minimizing impacts on local or regional resources.

11. Close interactions should be developed with materials suppliers, customers, and representatives of other industries, with the aim of developing cooperative ways of minimizing packaging and of recycling and reusing materials.

A major implication of these principles is that environmental concerns should be an integral part of the initial design activities for virtually every anthropogenic system. Thus, for example, concurrent process and product engineering activities in the manufacturing sectors should include all relevant environmental considerations. Similarly, engineers and architects should not design buildings or infrastructure components without relevant input regarding the environmental implications of, for example, their material choices, siting decisions, and mechanical and electrical system requirements. Crop selection and practices in agriculture and forestry should be appropriate to local conditions, require minimal input of scarce or environmentally problematic resources such as water or pesticides, and support long-term maintenance and enhancement of the pedosphere (soil layer). Concomitantly, therefore, it should be a policy goal to create incentives for and reward such behavior through, for example, procurement requirements and adoption of relevant standards.

4.5 PERSPECTIVES ON INDUSTRIAL ECOLOGY

Robert Socolow of Princeton University has identified six perspectives based on systems principles that are provided by industrial ecology:

1. Industrial ecology focuses on *long term habitability* rather than short term or ad hoc approaches, which tend to characterize current practice. The effort is to understand anthropocentric disruption to fundamental life-supporting systems and cycles rather than just responding to the obvious localized perturbations.

2. Industrial ecology focuses on concerns which are of *regional and global scope*, and persistent and difficult to manage.

3. Industrial ecology focuses on cases where the *scale of human activity overwhelms natural systems*.

4. Industrial ecology attempts to understand and protect the *resiliency of natural and human systems*, while identifying and minimizing impacts on more vulnerable systems.

5. Industrial ecology uses such systems techniques as *mass-flow analysis* to understand economic and environmental systems, and the linkages among them.

6. Industrial ecology views *economic production agents*—private firms—as central to mitigating environmental impact, and seeks to understand how their behavior might become more environmentally appropriate, rather than viewing them as blameworthy.

4.6 ILLUSTRATIVE CASE STUDY: THE AUTOMOTIVE TECHNOLOGY SYSTEM

Examples often make a difficult concept easier to understand. Accordingly, it is useful to use a common object, the automobile, and explore both the artifact and the sector behind it from an industrial ecology perspective.

4.6.1 Evolution of the Automobile as Artifact

An appropriate place to begin is to consider the apex of the initial stage of evolution of the post-World War II automobile technology. This arguably occurred in the late 1960s, when the most desirable automobiles were powered by what aficionados fondly called "Detroit iron": big V-8 engines that were relatively crude but effective. These so-called "muscle cars" consumed enormous amounts of gas, frequently getting less than 10 miles per gallon, and the untreated exhaust was high in hydrocarbon and NOx (nitrogen oxides) concentrations. But they were fast (at least in a straight line) and popular with customers; ever-larger engines became a significant competitive dimension in automobile marketing. Then came Earth Day in 1970, and the energy crisis of the early 1970s. Pollution control equipment was superimposed on existing engine designs. Demand for improved gas mileage increased, culminating in the establishment of corporate average fuel economy requirements in the Energy Policy and Conservation Act in 1975. Predictably, average engine size, efficiency, and power dropped as a consequence.

Yet the drop in automotive performance, measured along almost any parameter, was temporary, and a second post-war generation of car began its evolution under these significant, and superficially mutually exclusive, pressures for improvement in both environmental and energy efficiency. The average size of the engine in passenger cars did, indeed, drop and stay smaller, but engine horsepower began to rise as the engineering of the drive and engine systems became more sophisticated. Accordingly, the ratio of horsepower to engine displacement increased significantly, indicating more efficient operation (Figure 4.9). Indeed, between 1975 and 1991, the fuel economy of the average new car improved significantly, from 15.8 to 27.8 miles per gallon. At the same time, absolute performance of the product was increasing, as measured by acceleration times (Figure 4.10). In fact, a 1992 National Research Council report noted that the average passenger car horsepower-to-weight ratio, an important indicator of performance capability, was greater in 1992 than at any time since 1975. The modern automobile unquestionably provides more performance per unit resource (in this case, gasoline).

Moreover, the automobile in 1997 is considerably safer, handles better, lasts longer, and offers far more amenities—such as advanced sound systems, on-board diagnostics and climate control systems—than it did two decades ago. Impressively, these gains have been matched by similar increases in environmental efficiency. Since controls were introduced in 1968, VOC (volatile organic carbon) and carbon monox-

Automobile Engine Systems, 1975-1990

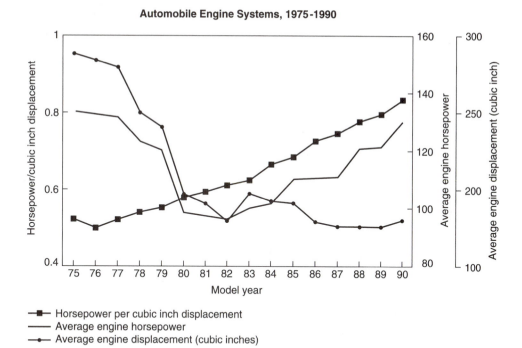

FIGURE 4.9   Despite smaller engines, average engine horsepower increased starting in 1982, and horsepower per cubic inch displacement grew steadily beginning in 1976.

ide emissions per vehicle have been reduced by some 96 percent, and, since imposition of NOx controls in 1972, emissions of those species have been reduced by over 75 percent.

In short, over the last two and a half decades, one of the principle—and defining —artifacts of the modern industrial economy has undergone an almost revolutionary change. It has improved its environmental performance on a per unit basis substantially; it is a far safer and more desirable product; and it has significantly enhanced not only its performance, but the efficiency with which it generates that performance.

4.6.2   Stages in the Integration of Environment and Technology

This evolution provides an interesting, if not perfect, analogy to the stages society must go through in its overall effort to integrate science, technology, and environmental considerations in all economic activity. The first stage of this integration process treats environmental impacts as overhead, as completely ancillary to the primary economic activity: fast cars or industrial production. The second stage begins to recognize that emissions must be controlled, but the underlying technological systems are not altered: this is the early 1970s car and the compliance stage of environmental regulation. In either case, the blend of naive "end-of-pipe" control and pre-existing technology is an uncomfortable one, and produces performance that is neither environmentally nor economically efficient. The next and more desirable stage, when environment becomes strategic, involves the re-engineering of the underlying technology to reintegrate eco-

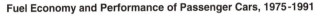

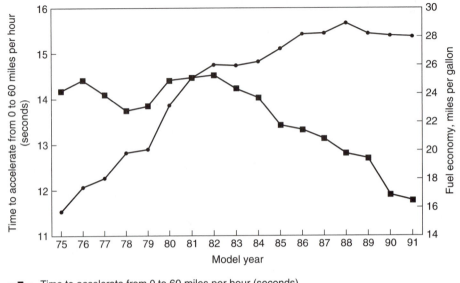

FIGURE 4.10   When combined with the performance data shown in Figure 4.9, the dramatic efficiency and environmental improvements in automobile engine function, resulting from the re-engineering of engine systems, are apparent.

Based on National Resources Council, Automotive Fuel Economy (Washington, D.C., National Academy Press: 1992).

nomic, engineering, and environmental efficiency from initial design through use and end-of-life resource recovery. This pattern is perhaps more universal than commonly recognized. In the replacement of solvents in manufacturing activities, for example, a similar three-stage approach has been taken (Table 4.2). Notice that only the last stage has the potential to produce significant integrated environmental and economic gains.

In this sense, then, the evolution of the automobile is somewhat analogous to the evolution of the environmentally efficient economy. The first stage is a Type I, linear economy that, just like a muscle car engine, takes in resources, uses them inefficiently, and pumps out substantial waste streams. The second stage is one where first generation end-of-pipe controls are placed on existing technologies, such as scrubbers on factories or emission controls on cars, which leads to a situation where neither environmental nor economic efficiency are optimized. This is, roughly speaking, the stage developed country economies are in now. The third stage is the fundamental systems redesign stage, which has been achieved for the automobile but not for any economies as a whole (although the product regulation focus in Europe, and much of the DFE and LCA work to date, is suggestive of such a systemic redesign).

What is somewhat ironic, of course, is that the automobile remains the most significant source of environmental impact in many areas of the globe. This results from non-technological factors. In many places, the benefits achieved by re-engineering the automobile as an artifact have been more than outweighed by population increases tied to higher per capita automobile ownership, combined with increased annual

TABLE 4.2   Integrating Environment into Technological Systems

Stage	Time to Implement	Primary Activities	Effects on Productivity	CFC/Chlorinated Solvent Example
1	Short term	Good housekeeping, emission-control equipment	None to negative	Replace tops on storage containers and install scrubbers
2	Medium term	Minor restructuring of existing processes	None to slightly positive	Eliminate unnecessary cleaning steps
3	Long term	Product, process, and operation redesign	Potential for substantial increases	Design products and processes to use aqueous solvent systems

The three stages of improvement of industrial cleaning processes, which resulted in a shift away from CFCs and chlorinated solvents to aqueous cleaning systems, are analogous to the automotive technology evolution discussed in the text.

mileage per vehicle, and a shift from fuel efficient sedans to four-wheel drive sports-utility vehicles. What technology has gained, changes in customer demand and vehicle use patterns, combined with population increases, have negated.

The automobile as an artifact, then, illustrates both the promise and the limitations of technological evolution in achieving sustainability. Its evolution is optimistic in suggesting that, if the technology systems handed down as a result of the previous stages of the Industrial Revolution are, indeed, re-engineered, substantial improvement in not just environmental performance, but also engineering and economic efficiency, might result. The viability of at least the "weak sustainable development model"—better quality-of-life with less environmental degradation—is supported by the automobile analogy. It suggests—although no persuasive data exist that come anywhere near proving—that substantial environmental improvement, at acceptable cost and offering increased quality-of-life, is possible with good engineering. At the same time, the evident limits to technological fixes are instructive: Consumers and society as a whole must not be left with the impression that simply relying on technology will avoid the need for difficult and complex political decisions. Better technology can buy time, but it cannot by itself buy sustainability.

It is precisely this non-technological dimension of the automobile that is so instructive to the industrial ecologist studying technology systems. The singular psychological and cultural appeal of the automobiles which is, unlike many artifacts, not limited to any region but is deep and global in scope, has no parallel in the modern economy. Any effort to understand the ecology of the automobile, or to regulate or modify its technological characteristics, fails unless this dimension of its existence is understood as a fundamental part of the automotive technology system.

4.6.3   The Automotive Technology System

It is obvious that, like any fundamental technological system, the automotive technology system is a complex system, and a rapidly evolving one at that. Taking a comprehensive overview of that system in terms of its most important levels, some of which are not usually considered when the environmental impacts of cars are contemplated, is thus a useful illustration of an industrial ecology approach.

Industrial Ecology Systems Hierarchies

The Automotive Technology System

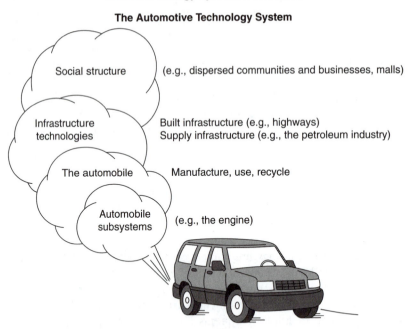

Social structure   (e.g., dispersed communities and businesses, malls)

Infrastructure   Built infrastructure (e.g., highways)
technologies   Supply infrastructure (e.g., the petroleum industry)

The automobile   Manufacture, use, recycle

Automobile   (e.g., the engine)
subsystems

FIGURE 4.11   The automotive technology system viewed from an industrial ecology perspective. Each level of the system must be evaluated not just by itself, but as it couples to other levels, if true economic and environmental efficiency gains are to be achieved.

Figure 4.11 is a simple but robust schematic of the automotive technology system. It includes the lowest, relatively technology-rich levels of a car's mechanical subsystems and the manufacturing processes by which it is made. These have been a predominant focus of environmental regulation, but, taken as a whole, they are probably not the major contributors to the automobile's environmental impact, except in a very local sense. Thus, for example, it makes sense to encourage the use of paints that do not contribute to local air pollution (e.g., low emissions of volatile organic compounds, or VOCs), but this is not a major environmental impact of the automobile.

The next level, automobile use, is more important. Moreover, there are two major dimensions to this system level: technical and cultural. On the technical side, great progress in reducing environmental impacts has been achieved, and more is possible. Examples yet to be implemented would include sensor systems that enable oil changes only when needed or report when air pressure in tires is low (low pressure results in lower gas mileage and greater tire wear). On the cultural side, however, failure to address environmental impacts is virtually total. The mix of cars purchased in developed countries, and soon to be available in major developing countries such as China, is increasingly inefficient. In the United States, families routinely purchase vehicles with four-wheel drive systems despite the fact that this functionality is almost never used, and the gas-saving federal 55 mile per hour speed limit was repealed in 1995 on the grounds it was an imposition on personal liberty and states rights. In Germany, attempts to impose any speed limits on the Autobahn routinely fail, even though the environmental party (the Greens) is widely popular. Even a cursory evaluation of this systems model, therefore, indicates that, in this instance, much attention probably is being focused on the wrong subsystem, and illustrates the fundamental truth that a strictly technological solution is unlikely to adequately address the cultural dimensions of such system dynamics.

On the other hand, for most of the world, the level of automobile recycling is one of the most successful aspects of the automotive technology system. In many developed countries, well over 90 percent of discarded vehicles are recovered for recycling, with over 75 percent of each recovered car by weight being recycled and economically reused. Moreover, this recycling system is truly a complex system, linked together by numerous price and cost feedback loops. The challenge at this level of the system is to not disrupt it. For example, changes in automobile composition that reduce the fractions of recoverable metal (steel or aluminum) below economic limits, or regulations that make such recovery uneconomic, could unintentionally destroy an environmentally important subsystem. It would be truly ironic, and a testament to the inability of most of society's existing institutions to think in systems terms, if this were to be done in the name of the environment.

Contrary to the usual understanding, the most significant environmental impacts of the automobile technology system probably arise from the next higher levels of the system: the infrastructure technologies and the social structure. Consider the energy and environmental impacts that result from just two of the major infrastructures required by the use of automobiles. The construction and maintenance of the built infrastructure—the roads and highways, the bridges and tunnels, the garages and parking lots—involves huge environmental impacts. The energy required to build and maintain such infrastructure; the natural areas that must be perturbed or destroyed in the process; the amount of materials, from aggregate to fill to asphalt, demanded—all of this is required by the automobile culture, and attributable to it. Similarly, a primary customer for the petroleum sector, and therefore, causative agent for much of its environmental impacts, is the automobile. Efforts are being made by a few leading infrastructure production firms such as Bechtel to reduce their environmental impacts, but these technological and management advances, desirable as they are, cannot in themselves compensate for the increased demand for infrastructure generated by the cultural patterns of automobile use.

The most fundamental effect of the automobile, however, may be in the geographical patterns of population distribution to which it has been a primary contributor. Particularly in lightly populated, developed countries such as the United States, the automobile has resulted in a diffuse pattern of residential and business development, which in turn now demands constant reliance on the automobile. Lack of sufficient population density along potential mass transit corridors makes such options uneconomic over much of the United States, even where absolute population density would seem to indicate otherwise (e.g., in densely populated suburban New Jersey). This pattern, once established, is highly resistant to change in the short term, if for no other reason than residences and commercial buildings last for decades. Thus, at this level of the system, high demand for personal transportation (i.e., the automobile) is firmly embedded in the physical structure of the community.

Figure 4.12 illustrates some of the relationships among cultural and technological aspects of the automotive technology system, and relates these to the time frame in

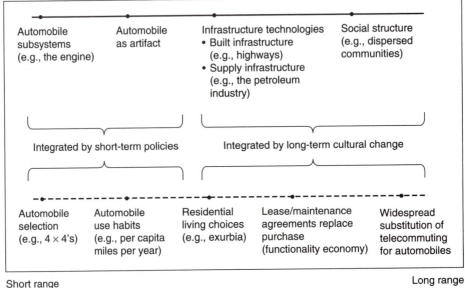

The Automotive Technology System:
Technology and Culture

Automobile subsystems (e.g., the engine)

Automobile as artifact

Infrastructure technologies
• Built infrastructure (e.g., highways)
• Supply infrastructure (e.g., the petroleum industry)

Social structure (e.g., dispersed communities)

Integrated by short-term policies

Integrated by long-term cultural change

Automobile selection (e.g., 4 × 4's)

Automobile use habits (e.g., per capita miles per year)

Residential living choices (e.g., exurbia)

Lease/maintenance agreements replace purchase (functionality economy)

Widespread substitution of telecommuting for automobiles

Short range

Long range

——— Technology
- - - Cultural

FIGURE 4.12   The interactions of technology and culture in the automotive technology system. Linking these systems, while difficult, can be achieved through the use of appropriate metric systems (see section 7.2).

which they can be evolved. Thus, for example, the engine is a principally technological system that can be, and has been, evolved relatively rapidly; the technology content of the infrastructure and the dispersed community structure, on the other hand, is relatively immune to evolution in the short term, principally because of the way it is embedded in the physical structure of the economy. Similarly, automobile selection can be evolved rather rapidly, even though it is a heavily cultural subsystem (after all, the shift to, and then away from, energy-efficient vehicles in developed countries has occurred within a 20 year time frame). On the other hand, development of teleworking alternatives to commuting in an automobile has been actively pushed by individuals and some governments for over 20 years as well, with relatively little to show for it. Here, the technologies to support this transition have long been available; they simply aren't being used.

Although brief, this discussion illustrates the value of understanding the system within which one is operating, and provides one possible way of understanding the automotive technology system. It also illustrates the fallacy, usually based on an incomplete understanding of the system, of focusing solely on technological subsystems when, in fact, the cause of poor systems performance lies in the cultural or economic realm. Even in such cases, however, technological evolution can serve an important role in mitigating environmental impacts over the short term even as changes in cultural patterns, which usually require much more time to evolve, occur.

REFERENCES

Allenby, B. R. and W. E. Cooper. "Understanding industrial ecology from a biological systems perspective". *Total Quality Environmental Management*. Spring 1994, 343–354.

Allenby, B. R. and D. J. Richards, eds. *The Greening of Industrial Ecosystems*. Washington, DC, National Academy Press: 1994.

American Electronics Association. *The Hows and Whys of Design for the Environment: A Primer for Members of the American Electronics Association*. 1993.

Ausubel, J.H. and J.E. Sladovich, eds. *Technology and Environment*. Washington, DC, National Academy Press, 1989.

Ayers, R.U. and U.E. Simonis, eds. *Industrial Metabolism*. Tokyo, United Nations University Press: 1994.

Cohen, J.E. "Population growth and earth's human carrying capacity." *Science*. 269: 341–346 1995.

Frosch, R.A. and N.E. Gallopoulos. "Strategies for manufacturing." *Scientific American*. 261(3):144–153 September 1989.

Graedel, T.E. "On the concept of industrial ecology." *Annual Review of Energy and the Environment*. 21, 1996.

Graedel, T.E. and B.R. Allenby. *Industrial Ecology*. Upper Saddle River, NJ, Prentice-Hall: 1995.

Graedel, T.E. and B.R. Allenby. *Industrial Ecology and the Automobile*. Upper Saddle River, NJ, Prentice-Hall: 1997.

Helm, J.L., ed. *Energy: Production, Consumption and Consequences*. Washington, DC, National Academy Press, 1990.

Lund University, Department of Industrial Environmental Economics. *Extended Producer Responsibility as a Strategy to Promote Cleaner Products*. 1992.

The Netherlands Department of Housing, Physical Planning and Environment. *National Environmental Policy Plan: To Choose or to Lose*. 1989.

The Netherlands Department of Housing, Physical Planning and Environment. *National Environmental Policy Plan Plus*. 1990.

The Netherlands Department of Housing, Physical Planning and Environment. *Integrated Substance Chain Management*. 1991.

Office of Technology Assessment. *Green Products by Design*. Washington, DC, US Government Printing Office, 1992.

Richards, D.J. and A.B. Fullerton, eds. *Industrial Ecology: U.S.-Japan Perspectives*. Washington, DC, National Academy Press, 1994.

Schmidheiny, S. (and the Business Council for Sustainable Development). *Changing Course*. Cambridge, MA, The MIT Press: 1992.

Schulze, P.C., ed. *Engineering Within Ecological Constraints*. Washington, DC, National Academy Press: 1996.

Socolow, R., C. Andrews, F. Berkhout, and V. Thomas, eds. *Industrial Ecology and Global Change*. Cambridge, England, Cambridge University Press: 1994.

Tilman, D., D. Wedlin, and J. Knops, "Productivity and sustainability influenced by biodiversity in grassland ecosystems." *Nature* 379:718–720 (and accompanying commentary by P. Kareiva, "Diversity and sustainability on the prairie." 673-674).

Turner, B. L. II, W. C. Clark, R. W. Kates, J. F. Richards, J. T. Mathews, and W. B. Meyers. *The Earth as Transformed by Human Action*. Cambridge, England, Cambridge University Press: 1990.

EXERCISES

1. Adopt and apply the industrial ecology principles listed in Section 4.4 to the agricultural sector.

   a. Based on your assessment, perform a gap analysis of that sector (that is, identify the greatest gaps between the desired state and current performance). What are the major gaps?

   b. Evaluate the ease with which the gaps you have identified can be closed. For which gaps, for example, can simple modifications and improvements of current practices result in substantial improvement, and which require fundamental advances in science and technology?

   c. Combine your gap analysis with your improvement analysis to prioritize the actions that could be taken to improve the environmental performance of the agricultural sector.

   d. Develop an integrated policy to achieve this.

2. Perform the assessments indicated in Question 1 for:

   a. The fisheries sector;

   b. The mining sector;

   c. The electronics manufacturing sector.

3. As discussed above, the automobile technology system has co-evolved with other sectors, including the petroleum sector.

   a. Discuss the implications of this co-evolution for the implementation of a car powered by hydrogen. What sectors would you expect to favor, and oppose, such an evolution? How could the predictable opposition to such an evolution be minimized?

   b. Under what circumstances should society pay firms or employees impacted by such technological evolution, and, if appropriate, how much should be paid?

CHAPTER 5

Industrial Ecology Infrastructure

5.1 DEFINITION

Many of the pioneering efforts to implement applications of industrial ecology, such as Design for Environment (DFE) methodologies in the electronics industry, integrated pest management (IPM) in agriculture, or life-cycle assessment (LCA) methodologies in the personal products industry, have been made by private firms. These efforts have clearly indicated that private entities, acting alone, can make some progress, but cannot compensate for wider, systems-based barriers to progress toward sustainability. The question therefore arises: What infrastructure must national-states and societies provide to enable individuals, private firms, and other entities to continue their progress in adopting practices and methodologies based on industrial ecology approaches? This enabling set of activities, which includes support of research and development programs, legal and economic reforms, and the development of appropriate educational tools, constitutes the industrial ecology infrastructure. Given the systems-based nature of the environmental challenge, this infrastructure will contain components at the local, national-state, and international levels.

The efficiency of a market economy, and its incentives for rapid innovation, are a necessary underpinning for the rapid technological evolution needed to progress toward sustainability. Even within a free market structure, however, there are four categories of activity that appropriately belong in the industrial ecology infrastructure, as opposed to being allocated to private entities to perform. These are (1) establishing general regulatory policies (boundary conditions) designed to encourage appropriate behavior on the part of producers and consumers; (2) establishing means to define and prioritize environmental risks, both among themselves and in conjunction with other risks and costs raised by policy decisions; (3) defining and prioritizing the values that private firms and individuals can use to make trade-offs in their operations; and (4) supporting research and development, and providing industrial ecology tools, which, because of their broad applicability and the inability of any private producer to capture the full benefits of the activity, are public goods.

5.2 ESTABLISHING APPROPRIATE REGULATORY POLICIES

The establishment of appropriate regulatory policies is discussed in Part II of this book. Here, the focus is on the need for, and appropriateness of, governmental action in this area.

Any economy is defined in large part through the formal and informal legal system that supports it and the culture within which it functions. Implementation of industrial ecology principles, and the required integration of environmental, technological, and economic activities that is implied by the shift from the "overhead" paradigm to the industrial ecology paradigm, constitutes a fundamental change in the economic structure. It is therefore not surprising that the implementation of such practices raises a number of basic legal and policy issues in areas apparently far removed from the environment, and, conversely, that many existing legal structures have at least the potential to affect private DFE activities. This is simply a reflection of environmental issues becoming strategic for society, and therefore requiring integration into the myriad of existing social, cultural, legal, and economic regimes that already exist.

In this context, it is clear that one of the most important things the government can do is facilitate the internalization of externalities through the use of fiscal tools such as fees, taxes, subsidies, and the like. Private firms can use any number of DFE and LCA tools, but they will only be able to function within the cost envelope created by a highly competitive global economy; they cannot, of themselves, price according to social cost rather than economic cost. Of course, national states are somewhat limited in this area as well. A country that establishes a high carbon tax (the benefits of which accrue to the world, not just the inhabitants of that country) might lose energy-intensive industry to other nation-states.

Another obvious issue is existing environmental laws, such as the Resource Conservation and Recovery Act in the United States, that presuppose linear manufacturing activities—a Type I system—and thus inhibit the development of desirable materials cycling patterns within the economy. An immediate effect of such laws is economically and environmentally unnecessary generation of residual material flows that must, for regulatory and economic reasons, be treated as wastes, and concomitant support of artificially high demand for virgin materials. Similarly, environmental laws that impose emissions control technology requirements not only impede evolution of more preferable control technologies, but, more subtly and insidiously, impede the evolution of more environmentally and economically efficient manufacturing technologies and product designs.

Other, less obvious examples include antitrust laws in general, which, regardless of national form, generally disfavor vertical integration and joint action among competitors regarding technologies and standards. Such laws can create difficulties where the environmental public policy is tending to encourage firm responsibility across the life cycle of its products and the constituent materials. Potential antitrust liability could arise, for example, should a large manufacturing firm in the United States try to take control of the secondary market in its refurbished products; or a consortium of firms representing significant market share select a few specific materials, or end-of-life materials recycling providers, in order to assure standardized environmental management. Demonstrating the need to be sensitive to national and cultural differences, it is likely that such antitrust considerations could be more important in the United States than in Europe, and in Europe as opposed to Japan, based on the different legal standards and assumptions about appropriate industrial organization that prevail.

An additional example is government procurement laws and regulations that frequently prohibit the purchase of refurbished or recycled products or products containing refurbished or recycled components or subassemblies, primarily to avoid vendor fraud. The reduction in incentives to environmentally appropriate behavior of potential suppliers that such requirements create are obvious.

In all of these cases, the issue is not that the policy goals behind the existing legal structures—control of inappropriate disposal of hazardous materials, avoidance of economic costs associated with monopolies or cartels, or reducing vendor fraud of public entities—are inappropriate. Rather, there is a need to revisit the way such goals are implemented in laws and regulations, which were often created before environmental quality was accepted as an important social good, to ensure that both the original policy ends and the new environmental ends can be accommodated. This policy accommodation process is, in fact, a logical facet of the evolution of environmental quality from an overhead to a strategic good for society.

It is equally clear that the task of integrating these different social goals and creating the regulatory structures, both environmental and non-environmental, that implement that integration are not appropriately delegated by society to private firms. Moreover, as with any system, it is desirable that policies be both appropriate to their level of the system's hierarchy—local issues should be addressed by local, not international, initiatives, for example—and that, taken as a whole, the integrated policy structure across all levels of the system be harmonized and reinforcing. This clearly implies a necessary role for international collaboration in developing this aspect of the industrial ecology infrastructure.

5.3 IDENTIFICATION AND PRIORITIZATION OF RISKS, COSTS, AND BENEFITS

Risk can be defined as the probability of suffering harm from a hazard. It is, however, one of those concepts that is intuitively obvious, and yet, perhaps for that reason, notoriously difficult to quantify. This is especially true when dealing with environmental issues and the inevitable trade-offs among environmental and other goods, where objective uncertainty is high and subjective values conflict. Assessing such trade-offs requires that the risks, costs, and benefits of not only the immediate subject, but the options and alternatives implied by that subject, be integrated into the decision-making process.

It is important for the industrial ecologist to understand these concepts, and for that reason they are dealt with in more detail in Chapter 10. Here, though, it is important to identify the two major categories of activities in this area that are appropriately part of the establishment of an industrial ecology infrastructure. The rationale for government involvement in each case is slightly different, yet in both cases compelling.

The first set of activities is to develop the scientific and technological understanding necessary to support the development of models and methodologies that can generate estimates of environmental risk in such a way that they can be efficiently reduced and compared with other risks. This involves, at the least, integration of traditional cost/benefit economic methodologies, systems-scale engineering risk assessments, and risk identification and assessment models that have been developed by environmental scientists. This is appropriately performed by society at large, as opposed to private interests, because such integrated models will inevitably require some decisions about values, and private interests will have at least potential biases that could skew the process toward protecting private interests at the cost of reducing social welfare. For example, few would trust risk models of smoking that are developed by the American Tobacco Institute.

5.4   PRIORITIZING VALUES

The second set of activities reflects the realization that, for society as a whole, risk inevitably contains a subjective as well as an objective component. Thus, identifying and prioritizing the values by which specific risks will, in their turn, be prioritized is a task that can only be done by society. Given different value systems, this task even theoretically cannot be accomplished unambiguously. Yet it must be done if there is to be any order in decisions chosen among options with different environmental and economic impacts. For example, current technology indicates that it is likely that any superconductor commercialized in the near term will contain toxic materials, and some have taken the position that no new use of toxics be permitted. The potential energy savings, which translate into less greenhouse gas emissions, could, however, be significant. Which should be preferred: less toxics or less greenhouse gas emissions? What model should be used to make such a comparison? Such an assessment involves only environmental risks; when concomitant, and very real, economic costs and benefits are integrated into the evaluation, it becomes complex indeed.

Such assessments may be complex, but they cannot be evaded. Private decisions are made tens of thousands of times each day, involve a multitude of known and unknown trade-offs among risks, and they impose the resultant environmental and economic costs, albeit in ignorance. Any hope of an absolute, objective, quantitative prioritization of environmental risks, both among themselves and in regard to other risks, costs, and benefits, must be regarded as unrealistic. It is possible, however, to do much better than has been achieved by current practices, both by generating information that reduces uncertainty and by achieving greater consensus on desirable endpoints of environmental policy.

There are two aspects of the consideration of risk issues that affect industrial ecology. The first is the requirement that comprehensive methods for evaluating all risks—environmental, economic, and otherwise—be developed. These risk issues, and the need to integrate risk assessment and cost/benefit methodologies, are discussed further in Chapter 10. A concomitant requirement, however, is that values be identified and prioritized to the extent possible, a task necessary for defining both the objective and normative dimensions of risks, costs, and benefits.

For example, if simple expenditure of money is taken as a criteria for social prioritization of risk (and it is not at all a bad one), then it is clear that, in the United States, by far the most important environmental risk is that of waste site remediation. Studies by the U.S. EPA and others, however, uniformly consider the environmental risks posed by such sites as minimal compared to other perturbations such as stratospheric ozone depletion, global climate change, and loss of biodiversity and critical ecosystems. Such disconnects also exist in other countries, reflecting to a large extent the fact that the shift of environmental issues from overhead to strategic is just beginning. In the short term, they are tolerable; in the longer term, resolution of these prioritization issues, which will require both substantial further research and political closure on fundamental questions of values and ethics, will be necessary if a common definition of a desirable sustainable world is to be reached. Otherwise, of course, resolution of situations where trade-offs are required becomes extremely difficult.

5.5 RESEARCH AND DEVELOPMENT

As regards research and development (R&D), it is apparent that a fundamental question is what the industrial ecology research and development (R&D) agenda should be; this issue is discussed in greater detail in Chapter 8. Equally as fundamental from a policy viewpoint is the recognition that, especially in this complex, multidisciplinary field, a preliminary sort must be made between those questions that require political answers—that is, are predominantly questions of value or subjective judgment—and those that are amenable to objective treatment. Take, for example, the issue of global climate change, which is currently quite contentious in countries such as the United States or the Persian Gulf states. Whether global climate change is occurring and, if so, what its impacts will be, is a researchable question. What should be done about it and who should bear the costs is a political issue that can be informed, but not resolved, by industrial ecology and scientific research. At present, the debate is clouded in part because, lacking data and understanding, both elements of the question are being answered by some based on ideology.

A second important issue regarding R&D is the question of where it is appropriately performed. Should a given research project be performed by the private sector, academia, the public sector, or some combination of the three? If, for example, a research project is not fairly directly linked to commercial advantage, it will probably not be performed by the private sector, which faces intense competitive pressures in virtually all industries. On the other hand, it is regarded by most people as inappropriate for the public sector to pay for research that benefits only a few firms and their shareholders. Lack of reasonably objective grounds on which to make such decisions, compounded by a lack of understanding of R&D, science, and technology by some political figures, has rendered this a somewhat ideological issue in many countries, particularly the United States.

REFERENCES

Allenby, B.R. "Supporting environmental quality: developing an infrastructure for design." *Total Quality Environmental Management*. Spring, 1993, pp. 303–308.

Science special issue on risk assessment, vol. 236, April 17, 1987.

U.S. Environmental Protection Agency. *Unfinished Business: A Comparative Assessment of Environmental Problems*. 1987.

World Resources Institute. *National Biodiversity Planning: Guidelines Based on Early Experiences Around the World*. Washington, DC, World Resources Institute: 1995.

EXERCISES

1. Assume that you are responsible for cleaning up a site where a variety of copper-containing compounds have been disposed of (remembering that copper is an aquatic toxicant). What kind of risk assessment would you perform to support your activity?

   a. Now assume that you are responsible for reducing the environmental impacts of automobiles in the United States. What kind of risk assessment would you perform to support your policy development activities?

   b. How do the two risk assessments differ?

   c. What are the potential economic impacts in each case, and how important do you think they are? Defend your answer.

   d. What are the potential technological impacts in each case, and how important do you think they are? Defend your answer.

2. In general, industrial ecology provides the scientific and technological basis for understanding the interrelationship between economic activity and underlying environmental systems, while the industrial ecology infrastructure involves the translation of this knowledge into methodologies and tools that can be put to use by individuals, firms, and governments in the short term. Using the example of materials, into which level do the following activities fall:

   a. Determining the aquatic toxicity of a particular lead-containing compound;

   b. Modeling the patterns of flow of cadmium in a developing country economy;

   c. Supporting a database that captures the environmental impacts embedded in common commercial plastics and feeds this information into industrial CAD/CAM (computer aided design/computer aided manufacturing) systems;

   d. Evaluating forest products to determine whether they are derived from forests that are being managed in an environmentally preferable way;

   e. Defining the parameters for a genetically engineered, environmentally preferable cotton suitable for growth in arid areas?

   Discuss and support your conclusions.

CHAPTER 6

Applications to Practice: Sector Initiatives

6.1 APPLICATIONS TO PRACTICE

In response to a number of pressures, including government regulation, customer demand, internal policies, and changing market conditions, a number of sectors have begun developing approaches to integrating the environment into their core activities, rather than treating it as overhead. In many cases, these initiatives, which are still mostly ad hoc and unsystematic, have been unintended, if beneficial, results of policies implemented for other reasons. It is obviously desirable to develop and deploy policies based on industrial ecology principles to forward these initial efforts.

Before building the future, however, it is important to understand the present, even if it is only in broad brushstrokes. Although society is only in the initial stages of the transition of environmental issues from overhead to strategic, with its increasing focus on implementation of environmentally preferable practices and technologies, it is already impossible to capture all developments in a single summary. Technological evolution, marginally but increasingly reflective of environmental objectives and constraints, is too complex and varied. This chapter, therefore, should be read as an introduction to the kinds of activities and technologies that are developing, rather than a rigorous discussion of developments in each sector. A longer reference section is provided at the end of this chapter for those seeking greater detail.

Nonetheless, there are notable commonalities across sectors, although specific applications to practice obviously differ. For example, primitive as they are, sectoral initiatives tend to have certain characteristics in common:

- They reflect a determination to take a more systems-based approach, even though in most cases the data are sparse and methodologies quite primitive.
- They adopt a life cycle approach to assessment of environmental impact.

- They build incrementally on existing practices, organizations, and cultures, rather than attempting radical, discontinuous change.
- They depend initially on individual champions, and in most cases are not yet institutionalized either within individual firms or within sectors.
- They rely heavily on experimentation and pilot projects; best practices or standard industry practices have not yet arisen, or, where they have, are still primitive.
- Important barriers to progress are not exclusively or even predominantly technological; in particular, economic mechanisms such as environmentally (and economically) inefficient subsidies are common and difficult to change.

A brief description of some of the current or easily foreseeable initiatives, outlined in Figure 6.1, provides some idea of the current state of development of the field. While use of the adjective "sustainable" is common in public discussion of these initiatives, and has been retained for clarity sake in this chapter, it must be remembered that this is short hand for "environmentally preferable." In no instance is it yet clear what truly sustainable performance would be.

6.2   DESIGN FOR ENVIRONMENT (DFE)

In the non-chemical manufacturing sectors, which include the manufacture of such varied articles as white goods (e.g., dishwashers, refrigerators, and clothes washers), electronics products, avionic systems and aircraft, and automobiles, implementation of industrial ecology principles is generally captured under the rubric of *DFE (Design for Environment) systems*. While there are issues specific to each sector—aircraft manufacture, for example, involves extensive use of carbon composites that are seldom encountered in electronics manufacturing—there are also significant common issues. For example, cleaning of metal piece parts and circuit board assembly are common

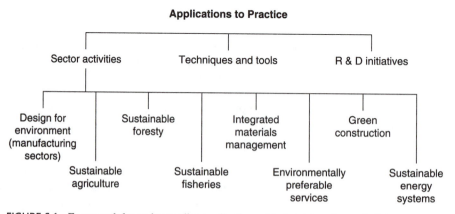

FIGURE 6.1   Framework for understanding applications of industrial ecology principles to current practices in various sectors.

manufacturing activities in many non-chemical manufacturing sectors. Therefore, joint DFE R&D is increasingly common, especially in the area of manufacturing processes. For example, AT&T and Ford have collaborated on a cleaning system that uses air-blasted frozen carbon dioxide pellet streams to clean circuit boards, thereby replacing chlorinated solvent baths.

There are two general categories of DFE. The first, *generic DFE*, represents actions that, if taken, improve the environmental performance of the affected entity as a whole. For a private firm, for example, generic DFE activities might include implementation of "green" management accounting systems, which break out from general overhead accounts the environmental costs associated with material, product, process, or technology choices, and thus encourage their minimization. It could also include development of environmentally sensitive procurement standards requiring, for example, that recycled material, rather than virgin material, be the default option for all specifications.

The second DFE category is *specific DFE*, which includes actions that impact specific product, process, and technology design choices. This might include development of specific computerized DFE design tools intended to be integrated into existing CAD/CAM systems. Such systems may be relatively simple extensions of existing design modules. For example, Design for Simplicity emphasizes using the least number of parts and materials possible, which is in most cases good environmental design as well. On the other hand, it may mean developing new dedicated DFE software, an activity aggressively being pursued at academic institutions such as MIT, Stanford, Berkeley, and the University of Tokyo; at industry consortia such as the Microelectronics and Computer Technology Corp. in Austin, Texas; or within individual companies such as IBM and Motorola.

6.3   SUSTAINABLE AGRICULTURE

Agriculture has significant environmental impacts. Land use is an obvious issue; Table 6.1 illustrates that significant areas of land have been degraded to lesser uses in the period between 1945 and 1990, primarily as a result of agricultural activities that have led to such impacts as nutrient loss, salinization (build up of toxic salts in the soil), compaction, and erosion. More than 85 percent of the wetland loss in the United States since the 1950s has occurred as a result of agricultural activities. Additionally, modern agriculture is heavily dependent on direct or embedded energy and chemical inputs. These, in turn, create significant infrastructure requirements involving their own environmental issues (e.g., manufacture of fertilizers, pesticides, and herbicides; production, shipment, and storage of petroleum products in underground tanks).

The obvious environmental costs associated with agriculture have led to a desire to explore the possibility of a "sustainable agriculture" system, defined by Pierre Crosson of Resources for the Future as "one that indefinitely meets demands for agricultural output at socially acceptable economic and environmental costs." Progress

TABLE 6.1   Global Anthropogenic Soil Degradation, 1945–1990

Degree	Definition	Area (M hectares)
Light	Mildly salinized, or eroded: widely spaced rills or hollows; > 70% of rangeland area still in native vegetation.	750
Moderate	Moderate erosion or salinization; soil chemical and physical integrity compromised; rangelands have 30-70% native vegetation.	910
Severe	Frequent gullies and hollows; crops grow poorly or not at all; rangeland has < 30% native vegetation.	300
Extreme	No crop growth occurs and restoration is impossible. May be caused by , for example, water erosion (Central Italy), wind erosion (Somalia), or salinization (former Soviet Union near Iran).	9
Total		1969

Source: World Resources Institute, *World Resources 1992-1993*, pp. 111–126. Oxford University Press, 1992.

toward more sustainable agricultural production—a term including but more comprehensive than existing practices such as "low-input," "organic," or "regenerative" farming—involves at least four initiatives: (1) integrated pest management, known as IPM; (2) sustainable tillage and land management practices; (3) sustainable crop selection; and (4) economic reform.

IPM replaces the total reliance on chemical pesticides with an integrated pest management system. Components of such practices might include release of natural predators of, and parasites on, the pest species; development of resistant crop species, either through genetic engineering or traditional plant breeding means; and substitution of targeted applications of chemical pesticides for blanket application. IPM generally does not aim for 100 percent elimination of crop loss, but at maintaining a sustainable high yield while tolerating some crop damage at the margin.

Sustainable tillage and land management practices include a number of possible options that are generally selected based on the characteristics of the specific site. For example, sustainable tillage patterns would vary considerably between a developed wheat growing area in the United States and Canada, and a rice growing area in Japan consisting of many small paddies scattered throughout the landscape. Examples of such practices might include temporal intercropping to avoid depleting the soil of specific nutrients, which frequently occurs when the same crop is planted in the same area (a pattern that is encouraged by the structure of some countries' agricultural subsidy programs), or spatial intercropping, which reduces the risk of loss inherent in wide-area monocropping. In the latter case, a single disease or parasite can more easily attack large areas.

Sustainable crop selection is an obvious means by which global agriculture can be made more environmentally efficient. Since prehistory, agriculture has tended to modify both the environment and plant species to produce desired crops through such means as irrigation and breeding programs. This has led to a little recognized but extraordinary reliance on an extremely limited number of species: Only seven crops—wheat, rice, corn, potatoes, barley, cassava, and sorghum—account for over 75 percent of the world's food supply, and the first four account for some 60 percent. This becomes undesirable if the environmental costs of such reliance, in terms of, for example, soil erosion, depletion of fossil groundwater resources, or soil salinization and waterlogging, become unsustainable, especially if there are alternative crop species better acclimated to local conditions. Among such species currently under active investigation are the grains quinoa, amaranth, triticale, millet, and buckwheat, and the winged, rice, fava, and adzuki beans.

Economic reform seems a strange inclusion here, yet it is critical. Perhaps no other sector of most national economies is so subsidized or otherwise distorted as the agricultural sector, either directly or through land use programs, in which farmers may be paid, for example, not to grow crops on certain lands. In 1994, overall subsidy payments to farmers amounted to $80.5 billion for the European Union, $46.4 billion for Japan, and $22.0 billion for the United States. Approximately half of the European Union's 1994 budget (49.3 percent) went to agricultural programs. Moreover, commodity support programs, a common subsidy mechanism, have the effect of limiting considerably the flexibility of farmers to decide what crops they want to plant, and are usually quite targeted. In the United States, for example, farm price support programs are limited to seven commodities: corn, wheat, sorghum, barley, oats, rice, and cotton. Moreover, because in the U.S. program price supports are paid only on acreage that historically has been dedicated to that crop, and it is forbidden to plant one crop on another crop's declared acreage ("crop base"), the effect is to propagate historical crop selection patterns regardless of changing environmental conditions. A recent World Resources Institute study by Paul Faeth states the case regarding the U.S. concisely, but his words have global applicability:

> We can change farm policies to save money and protect the environment. We can achieve these goals while protecting farm income. All we have to do is reform the rigid and costly subsidy programs that have long dominated U.S. farm policies.

Another obvious economic distortion associated with agricultural activities is the subsidization of infrastructure, especially irrigation infrastructure in arid areas such as the Western United States, and in countries where farmers are politically powerful, such as India. Legal systems and politics make such subsidies difficult to dismantle once they are granted. U.S. water allocation law, for example, exacerbates water subsidy distortions in the West by requiring that farmers use their full allocation of water, derived from historic use patterns, or lose it going forward. The result is that water waste may be mandated to preserve future rights to water resources, even as residential or industrial users are struggling with a drought and willing to pay significantly more for what is, for them, a limited resource. In India, political parties have traditionally sought the rural vote by competing with each other to promise lower irrigation water prices.

6.4   SUSTAINABLE FORESTRY

Sustainable forestry is similar to sustainable agriculture in many ways in that technologies are being evolved to continue production of desired species with minimal inputs of energy and chemical materials and less impact on surrounding systems (e.g., ground and surface water systems). A major difference in definition of sustainability arises, however, from the similarity between forestry and agriculture: raising one or a limited number of species of interest. Agriculture, after all, is not intended to replicate nature, while a natural forest is, in fact, a complex biological community. The term "sustainable forestry" is meant by some to mean a process in which, for example, a tree is planted for every tree harvested, but such wooded areas can come to represent the same monoculture patterns as traditional agriculture.

It is arguable that the term "sustainable forestry" should be reserved for forested areas that are sustainably used and harvested for a variety of ends—nuts, eco-tourism, wood—but that do not in the process lose their complex community structure, or in those cases where an existing forest community is being exploited sustainably without degradation. On the other hand, "sustainable agriculture" more accurately describes the monoculture of trees for intensive exploitation, with concomitant loss of biological diversity compared to a forest ecology. From an industrial ecology perspective, one of the important justifications for this approach is that a "sustainable agriculture" system, whether it be crops or trees, is generally a more intensively engineered and managed system than a "sustainable forestry" operation.

Obviously, the opposite of sustainable forestry is deforestation, which, as in the case of sustainable agriculture, frequently is a result of distortions introduced by inappropriate economic incentives. As Rice et al. note, "national agricultural policies, road development and colonization can each pose a far greater danger to tropical forests than unsustainable logging." For example, a major cause of deforestation in the Brazilian rainforest was a tax subsidy encouraging conversion of the land to pasture. Moreover, roads built into tropical forests to support activities other than logging, such as petroleum production, frequently produce concomitant new settlement and local deforestation. A major cause of deforestation in the U.S. Pacific Northwest is below-market value sale of trees on U.S. public land to lumber companies, a large if hidden subsidy with substantial negative environmental externalities.

Partially as a result of such differential policies, partially as a result of differences in technological and economic behavior, and partially as a result of economic development status and land allocation patterns, patterns of forestation and deforestation tend to vary widely among countries. Table 6.2, for example, shows those countries in which annual deforestation rates in the 1980s exceeded 2 percent of remaining reserves, and those countries where the reforestation rates during that decade were 0.5 percent or above. Interestingly enough, as Table 6.2 also shows, the most significant deforestation in terms of area are not in the same countries that show the highest rates of deforestation. It is less surprising that high rates of deforestation occurred in Asian and Latin American countries, while reforestation occurred in developed countries.

6.5   SUSTAINABLE FISHERIES

Tripling from 1958 levels, the world catch of marine fish peaked in 1989 at 86.4 million metric tons (Figure 6.2), and has since decreased to a little over 80 million metric tons despite (actually, because of) continuing investment in new fishing technology. From 1972 to 1992, for example, the number of fishing boats doubled, going from about 580,000 to 1.2 million, but, according to the United Nations Food and Agriculture Organization, this was a poor investment: Some 124 billion dollars per year of invest-

TABLE 6.2   Change in Forest Resources*

Country	Annual Deforestation % average 1981-1990	Annual Deforestation 000 hectare 1981-1990	Annual Forestation % average 1981-1990
Bangladesh	3.3		
Costa Rica	2.6		
Dominican Republic	2.5		
El Salvador	2.1		
Haiti	3.9		
Jamaica	5.3		
Pakistan	2.9		
Paraguay	2.4		
Philippines	2.9		
Brazil		3,671	
Indonesia		1,212	
Zaire		732	
Mexico		678	
Bolivia		625	
Venezuela		599	
Thailand	2.9	515	
Germany			0.5
Hungary			0.5
Ireland			1.3
Portugal			0.5
Switzerland			0.6
United Kingdom			1.1

Note the inverse relationship between development status and deforestation activities.
Data from World Resources, 1994-1995 (Oxford, Oxford University Press, 1995)

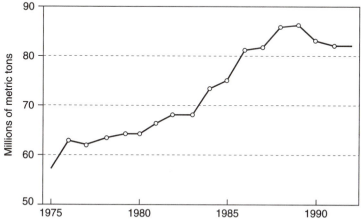

Yearly Catch of Marine Fish, 1975-1992

FIGURE 6.2   Yearly Catch of Marine Fish, 1975–1992
Based on UNFAO data

ment obtained about 70 billion dollars per year in revenue. Moreover, harvesting technology—from sonar systems that detect schools of fish, to larger gill nets, purse seines, and drift nets—has gotten more effective, so that each boat can potentially catch far more fish than the older, less technologically advanced fishing boats. In addition, because of the indiscriminate nature of these systems, the volume of "bycatch"—fish that are too small, or noncommercial species of all kinds, which are generally simply thrown overboard—has increased substantially as well. In some cases, the fraction of bycatch is far greater than that which is kept. Trawls and dredges by design impact the ocean bottom, with significant negative impact on benthic species such as oysters.

Not surprisingly, a number of fisheries, including those of the Pacific anchovy, the Atlantic cod, the haddock, the cape hake, the southern bluefin tuna, and the chub mackerel, have collapsed or are declining. Not all of this trend, however, can be attributed to advanced fishing technologies: The estuarine and coastal environments that act as nurseries for many species are being both developed, and polluted by hard-to-control, non-point sources such as upstream agriculture.

In large part, the fisheries story is a classic example of "the tragedy of the commons." Because fish cannot be reduced to private property until landed, each fisherman has a strong incentive to exploit the resource past sustainable limits: The cost of a declining fishery is borne by all, while the benefit of his overfishing accrues to him directly. Ironically, the breakdown in fisheries has for the most part occurred not as a result of increased activity by "traditional" fishermen, but as a result of new industrial-scale fishing technologies, which are much more efficient—and expensive. The need to earn a return on this new capital has driven the over-exploitation and subsequent collapse of fisheries. This, in turn, drives even traditional fishermen to extreme means, such as the practice of using dynamite to kill reef fish, which, of course, also destroys the reef.

One solution is to create a system in which the resource is "owned" by individuals who have the right to exploit it over time, and thus the incentive to protect its long-term viability. While the political problems with this approach are numerous—subsidizing uneconomic fishing infrastructures is far more politically viable than creating and allocating such rights, even if fisheries do degrade over time as a result—one country, New Zealand, has actually tried it.

New Zealand established in 1986 an "individual transferable quota," (ITQ) system, which granted the right to harvest a given amount of fish in the form of a private property right, namely, the ITQ. The ITQ can be used directly, traded, or sold; the total number of ITQs per species is determined by establishing a "total allowable catch," or TAC, for that type of fish. The system is explicitly intended to reduce not just overfishing, but overcapitalization of the sector as well. There are several problems with implementing such systems more broadly, including lack of data about the success of the New Zealand approach, a clash between ITQ and the traditional "winner-take-all" fishing culture, the difficulty of establishing TAC, and the complexity of implementing ITQs for species that cross international boundaries. Nonetheless, the New Zealand approach is promising.

6.6 INTEGRATED MATERIALS MANAGEMENT

The importance of understanding materials, their energetic and environmental impacts, and their cycles is both intuitively apparent and an important element of the practice of industrial ecology to date. For most people, materials have about them an air of transcendental permanence. Materials selection and use are, however, profoundly cultural phenomena, and the nature of material use and the psychology of materials have shifted significantly in the past half century, with profound environmental implications. In developed economies, anthropogenic organics have replaced "natural" materials to an extent that constitutes a difference in kind, not just degree. It is thus worth starting this section with some observations by Jeffrey Meikle comparing the pre-World War II world of wood and iron with the post-War expansion of the plastics industry:

> The utopian impulse of the 1930s had indeed striven for static perfection, but that of the 1950s abandoned perfection in pursuit of an ever-improving flux. . . . The very extravagance of artificial forms depended on synthetic materials that yielded easily to human manipulation. . . . Whether they [design critics] criticized Americans for splurging on "borax and chrome" or became self-consciously involved in "learning from Las Vegas," they confronted a vigorous popular culture that owed much of its exuberant surrealism to plastic. . . . Without anyone realizing it, Thomas Hine has observed, 'the very nature of *things* had changed' as people adjusted to 'a disposable world.' In so doing, they lost control of the process in two crucial but contradictory ways. Neither became clear until much later. In the first place, as material things became attenuated, images or simulations of things assumed a cultural significance nearly equal to that of things themselves. At the same time, however, the flood of attenuated things swelled to such proportions that no matter how disposable each individual thing might be, taken together they threatened to swamp landfills and discharge chemical waste into the air and water on which life depended. Plastic offered an unprecedented degree of control over individual things—allowing extravagance of form to coexist with precisely engineered function. But plastic also accelerated larger processes that society recognized as out of control only long after it had become dependent on the comfort and convenience of plastic.

One might almost postulate a materials progression leading to increased dematerialization of economic activity: from use of traditional, long-lasting materials such as metals and wood to the plastic-oriented, disposable consumer society to the information society, where information and intellectual capital increasingly displace material and energy-intensive technologies. While this schema remains speculative at this point, it is important to recognize the fundamental truth that material use and selection, and concomitant environmental impacts, are a profoundly cultural phenomenon.

Perhaps for this reason, anthropogenic material flows and impacts on natural material cycles are not well understood by many people. For example, sensitized by concerns over hazardous materials, most people are unaware of the fact that even in developed countries by far the greatest component of materials consumption is simply sand, gravel, and aggregate—common construction minerals. Figure 6.3, based on Bureau of Mines data for the United States, illustrates this point, as well as the fact that even in 1990, recycling efforts in the economy as a whole were relatively successful (approximately 10 percent), especially so for metals as a group (over 50 percent). Such recycling patterns are important in developed countries because of the disparity in material use between them and developing countries (Figure 6.4). Such disparities also exist to a lesser extent among developed countries (Figure 6.5), reflecting cultural, historical, ideological, and physical factors. As Table 3.3 illustrates, it would take a sub-

Material Use by Commodity Groups
Million metric tons (1990)

Apparent consumption = 2,533

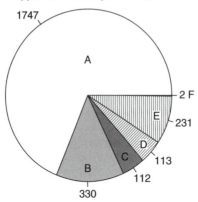

Recycled quantity = 245

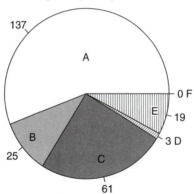

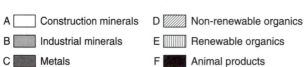

A ☐ Construction minerals D ▨ Non-renewable organics

B ▨ Industrial minerals E ▦ Renewable organics

C ■ Metals F ■ Animal products

FIGURE 6.3 Material Use by Commodity Groups. Note particularly the high recycling rate for metals.
Source: U.S. Bureau of Mines

stantial increase in world production of many basic commodities for world per capita material consumption to come up to the levels of consumption in the United States.

An essential element of industrial ecology is the concept of closing material loops where environmentally appropriate. This has the important benefit of slowing the velocity of materials through the economy (time from initial production to ultimate disposal), which, in turn, means that a higher quality of life for more people can be supported on the same size resource base. This implies, however, that an explicit systems-based approach to material utilization must be adopted. This is in contradistinction to traditional practice, in which no single economic actor is responsible for more than a segment of the life cycle of

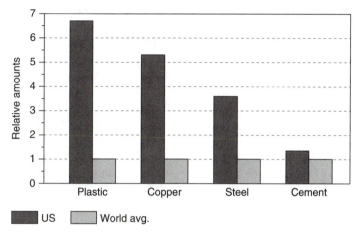

**Relative Materials Use
per capita**

☐ US   ☐ World avg.

FIGURE 6.4   U.S. consumption normalized to world average. The environmental degradation which would be associated with global material consumption patterns similar to those of the U.S. today would be many times that currently experienced.
   Source: U.S. Bureau of Mines data.

**Developed Country Material Patterns
(figures are metric tons per year, 1991 data)**

	USA	Japan	Germany	Netherlands
Commodities consumed	5149	2133	1773	574
Domestic to foreign commodities ratio	8.1	2	3.4	.9
Commodities/capita	20	17	22	38
Secondary material consumption[1]	16,088	3583	4991	701
Ratio of commodity to secondary material consumption	3	2	3	1
Total material requirement per capita	84	46	86	84

[1]Material that is disturbed by economic activity, but which does not enter the economic system, such as mining overburden

FIGURE 6.5   Some differences in patterns of materials consumption among developed countries are quite striking. Note, for example, the differences between the U.S. ratio of domestic to foreign commodities, and that of the Netherlands, which must import most of its materials. The low total material consumed per capita in Japan, as opposed to the other countries listed, is also striking.
   Source: Data from World Resources Institute, Resource Flows: The Material Basis of Industrial Economies, 1997

any material. The few exceptions have been sporadic and of limited duration. For example, some large manufacturing firms integrated vertically into materials production, such as automobile manufacturers that, for a time, attempted to mine and process their own iron ore, while other firms, such as the Bell Telephone System in the United States, managed materials derived from their recycled leased products. Concomitantly, the materials flow associated with industrial economies tended to be linear—from virgin material, to use, to disposal—and such recycling as did occur was underfunded. Moreover, reflecting the minimal economic incentives to recycle most materials, virtually no significant R&D resources were targeted at recycling processes or products, and the technologies associated with recycling have tended to be primitive in contradistinction to manufacturing technologies.

The beginnings of the transition that will change traditional materials supplier firms—such as aluminum and other metal producers, plastics manufacturers, and chemical manufacturers—into full service material management firms are already visible. The U.S. chemical industry has developed the Responsible Care program, which not only calls for environmentally responsible manufacturing, but also helping customers use the resulting products in a responsible manner. Plastics manufacturers are developing post-consumer takeback programs and devoting increased amounts of their R&D budgets to finding uses for recycled polymer. Recycling of metals, which has always occurred at some level, is an increasingly important consideration in designing new technologies and related policies: A major consideration of automobile takeback systems, for example, is that they not skew the economics away from existing practices that recycle virtually all of the steel in a discarded automobile.

An important policy implication of this evolution in materials management should be recognized. An important driver of many environmental initiatives regarding residual materials has been the occasionally irresponsible way they have been managed in the past. Many sites in the U.S. Superfund program, for example, are locations where recycling of used motor oil, lead batteries, or other hazardous materials occurred historically. The regulatory response has been to heavily regulate residual flows of any kind (at one point, the U.S. Environmental Protection Agency even proposed to regulate any returned product, such as a telephone or automobile, as a hazardous waste, which would have immediately destroyed virtually all of the product takeback and end-of-life product management systems that did exist). This pattern is evident in the Basel Convention, which limits international flow of hazardous residuals, and the U.S. Resource Conservation and Recovery Act. Materials so regulated cannot compete with virgin materials, so, in a form of self-fulfilling prophecy, being listed as waste under such programs ensures that the materials will, indeed, become waste and not be reused. In short, a linear flow of materials throughout the economy is, ironically, being reinforced by environmental regulation.

What integrated materials management means is that the historic pattern, in which production of virgin materials was treated as of high economic value but materials recycling was regarded as marginal at best (and hence gave rise to significant environmental impacts), is changing. The large chemical, metals, and plastics firms that produce virgin materials are now beginning to focus on life-cycle management, including recycling technologies and post-consumer residual management. The environmental costs of improper end-of-life handling of materials are being driven into the demand structure, with the result that suppliers cannot ignore them any longer. Suppliers of polyvinyl chloride (PVC) plastics, for example, will see their demand fall to the extent they cannot ensure recycling systems that do not generate dioxin emissions as a byproduct. At the limit, manufacturers will "rent" materials from their suppliers to be incorporated into their products with the material management firm retaining ownership and responsibility for the same material over a number of product lives. Dow Chemical Company, in fact, already offers some of its solvent customers such a "material lease" program.

Accordingly, the policy challenge is to move toward a regulatory regime that recognizes the value of and encourages cycling materials within the economy, while at the same time providing assurance that handling residuals in the global economy does not lead to any unnecessary risk because of improper, or "sham," recycling. The purpose of an integrated material policy should be to ensure safe handling of all materials regardless of where they are in their life cycle, and to regulate every material equally, regardless of whether it is virgin or originates in a residual stream, according to the actual risks involved.

The need for economic reform in this sector also requires mention, primarily because of the extraordinary subsidies that are sometimes involved. These subsidies can go far beyond the usual favorable tax treatment mechanisms, such as depletion allowances, to other direct and indirect supports. For example, the U.S. Mining Law of 1872, which still controls mineral rights on public land, is predicated on an assumption that mining is the highest use to which any public land can be put, which leads to such extraordinary results as the acquisition of 1,800 acres in Nevada in May 1994 by a Canadian company, which paid the U.S. government $9,000.00 for gold resources valued at 8 to 10 billion dollars. Yet another hidden subsidy is the failure of many countries, including the United States, to insist that mining companies remediate the environmental damage resulting from their operations, although this is beginning to change. Indirectly, the subsidies to the energy sector also subsidize the production of many virgin materials because their production is relatively energy intensive.

6.7   ENVIRONMENTALLY PREFERABLE SERVICES (EPS)

As Table 1.1 suggests, traditional environmental policies have focused primarily on specific materials of concern and the manufacturing sector as the "point source" of emissions, while services have generally been ignored. An industrial ecology perspective presents a quite different picture: Services are a key component of a sustainable economy. Service sectors are critical both in driving the evolution of environmentally preferable technologies and practices (EPS as customers), and in providing environmentally preferable alternatives for existing activities (EPS as providers). Services such as transportation can also be an unrealized source of substantial environmental impacts in a global economy.

Services are an increasing component of every developed national economy. In 1993 in the United States, for example, 79 percent of employment and 76 percent of economic value-added was in the services sector, which extends a decades old shift from agriculture and manufacturing to services (Figure 6.6). Services are not, however, a homogeneous sector, nor are they well-defined: The category includes, for example, transportation, communications, health care, retailing, and financial activities, all of which have very different environmental implications. Thus, Table 6.3 provides an overview of the economic divisions recognized by the U.S. Standard Industrial Classification (SIC) codes; Divisions E through J arguably reflect services. In Table 6.4, the wide variety of activities included under "services" is apparent, and argues for more sophisticated evaluation of the environmental impacts of services than is usually the case. The environmental impacts of hospitals, the U.S. Postal Service, retailers, law firms, and phone companies are obviously dissimilar, and need to be evaluated on their own merits.

While this economic evolution has been thoroughly discussed in the literature, its implications for environmental policy have not been well understood. Services are, however, significant in four principle ways:

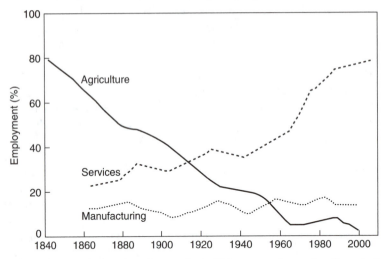

Evolution to Service Economy: Employment Trends

FIGURE 6.6   Evolution of employment in the U.S. economy from primarily agricultural to predominantly services. Similar trends mark other developed economies, although the percent employed in services is usually less than that in the U.S.

TABLE 6.3   Major SIC Code Divisions

Division A.	Agriculture, forestry and fishing
Division B.	Mining
Division C.	Construction
Division D.	Manufacturing
Division E.	Transportation, communications, electric, gas, and sanitary services
Division F.	Wholesale trade
Division G.	Retail trade
Division H.	Finance, insurance, and real estate
Division I.	Services
Division J.	Public administration
Division K.	Nonclassifiable establishments

1. Service firms, particularly large retailers, can exercise significant leverage over the environmental performance of suppliers because of their large size and capability to aggregate demand from many small users.

2. Service firms, which are frequently the closest to the end customer, can play an important role in educating the consumer (for example, Home Depot, a large retailer of construction and home maintenance supplies, is a leader in supporting and publicizing life cycle assessments studies of such products).

3. Services can facilitate environmentally preferable patterns of resource and product use (for example, energy management services, increasingly offered by energy utilities as part of their demand-side management (DSM) programs, help customers reduce energy use and, thus, reduce the need to build new generating capacity as well).

TABLE 6.4   Major Group Categories in SIC Code Divisions

Division E. Transportation, communications, electric, gas, and sanitary services

Major Group 40.	Railroad transportation
Major Group 41.	Local and suburban transit and interurban highway passenger transportation
Major Group 42.	Motor freight transportation and warehousing
Major Group 43.	United States Postal Service
Major Group 44.	Water transportation
Major Group 45.	Transportation by air
Major Group 46.	Pipelines, except natural gas
Major Group 47.	Transportation services
Major Group 48.	Communications
Major Group 49.	Electric, gas, and sanitary services

Division F. Wholesale trade

Major Group 50.	Wholesale trade–durable goods
Major Group 51.	Wholesale trade–nondurable goods

Division G. Retail trade

Major Group 52.	Building materials, hardware, garden supply, and mobile home
Major Group 53.	General merchandise stores
Major Group 54.	Food stores
Major Group 55.	Automotive dealers and gasoline service stations
Major Group 56.	Apparel and accessory stores
Major Group 57.	Home furniture, furnishings, and equipment stores
Major Group 58.	Eating and drinking places
Major Group 59.	Miscellaneous retail

Division H. Finance, insurance, and real estate

Major Group 60.	Depository institutions
Major Group 61.	Nondepository credit institutions
Major Group 62.	Security and commodity brokers, dealers, exchanges, and services
Major Group 63.	Insurance carriers
Major Group 64.	Insurance agents, brokers, and services
Major Group 65.	Real estate
Major Group 67.	Holding and other investment offices

Division I. Services

Major Group 70.	Hotels, rooming houses, camps, and other lodging places
Major Group 72.	Personal services
Major Group 73.	Business services
Major Group 75.	Automotive repair, services, and parking
Major Group 76.	Miscellaneous repair services
Major Group 78.	Motion pictures
Major Group 79.	Amusement and recreation services
Major Group 80.	Health services
Major Group 81.	Legal services
Major Group 82.	Educational services
Major Group 83.	Social services
Major Group 84.	Museums, art galleries, and botanical and zoological gardens
Major Group 86.	Membership organizations
Major Group 87.	Engineering, accounting, research, management, and related services
Major Group 88.	Private households
Major Group 89.	Miscellaneous services

Continues

TABLE 6.4   Continued.

Division J.	Public administration
Major Group 91.	Executive, legislative, and general government, except finance
Major Group 92.	Justice, public order, and safety
Major Group 94.	Administration of human resource programs
Major Group 95.	Administration of environmental quality and housing programs
Major Group 96.	Administration of economic programs
Major Group 97.	National security and international affairs

TABLE 6.5   Consumer Products and Healthcare Sales (1993; Figures in Billions of U.S. Dollars)

Industry	Product companies		Service providers	
Toys	Hasbro	$2.7	Toys 'R' Us	$7.9
	Mattel	2.7		
Healthcare	Johnson & Johnson	$14.1	McKesson Corp.	$11.7
	Bristol-Myers Squibb Co.	11.4	Walgreen Company	8.3
	Merck	10.5	Bergen Brunswig	6.8
	Baxter international	8.8	Hospital Corp. of America	5.1
	American Home Products	8.3	Humana, Inc.	3.1
Food and general merchandise	Procter & Gamble Co.	$30.4	Kroger Company	$22.3
	Philip Morris (Food)	34.5	American Stores Company	18.8
	PepsiCo	25.0	Safeway, Inc.	15.2
	RJR Nabisco Holding Co.	15.1	Fleming Corp.	13.1
	Levi Strauss	5.8	Wal-Mart Stores, Inc.	67.3
	Burlington Industries	2.1	Sears, Roebuck & Company (merchandise division)	29.6
	Spring Industries, Inc.	2.0	Kmart Corp.	34.2

4. Services can be direct substitutes for energy and material use in some cases (telecommuting being the most familiar example).

Providers of services are frequently substantial consumers of the underlying product. For example, as shown in Table 6.5, large retailers are frequently of the same magnitude as the supplying manufacturing firm; accordingly, they can substantially impact the environmental performance of their suppliers. Moreover, when large service providers work with their suppliers to achieve better environmental performance, they are in essence fulfilling the mantra of environmental policymakers: They are acting to internalize environmental externalities. Banks and financing companies do the same thing when they condition credit on the environmental performance of a borrower, either specifically (by requiring that the industrial site offered as collateral for the loan be clean) or generally (environmentally superior performance is seen as an important indicia of the risk levels, and thus general creditworthiness, of a firm). What was previously not captured in the economic system—or else, after all, there would be no need for the service firm to work with its supplier—is converted by the service firm into a component of a financial transaction: It is "monetized."

6.7.1   The Information Revolution

As has been stated elsewhere in this text, it is a strong hypothesis that greater environmental efficiency, and the evolution of a more sustainable global economy, will require at least partial substitution of information and intellectual capital for other economic inputs, most obviously materials and energy. This particular aspect of environmentally preferable services therefore deserves a little more attention.

The evolution of the information industry, including both the electronics manufacturing and information services sectors, suggests that this trend may not only be occurring, but may be economically robust. The exponential growth in the performance of information processing technology is well documented; Figure 6.7 shows the growth in bits that can be stored per memory chip, while Figure 6.8 shows the trend in central processing unit performance for four different systems, from microprocessors to supercomputers. As Figure 6.9 demonstrates, this increasing computational power has been accompanied by reduced use of materials and energy, with lower prices, and increased economic efficiency (lower cost per unit of performance, in this case measured in MIPs, or million instructions per second). Although the data are hard to evaluate, these cost and efficiency trends are apparently contributing to substantial growth in "information stocks"; Figure 6.10 presents such data from Japan. Consumption of information platforms, from transmission capability such as fiber optic cable and spectrum space to CD-ROMs and cellular telephones, is also increasing rapidly.

Overall, then, the impression is of an information sector with rapidly evolving technology and falling costs per unit performance. This would appear to contradict

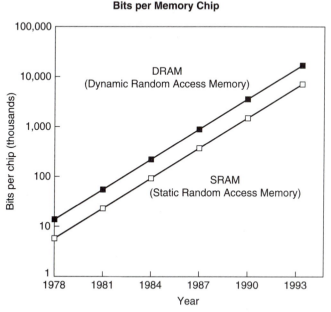

FIGURE 6.7   Growth trends in memory chip capacity. Note that this is a semi-log figure, so that the increases in memory capability are actually exponential.
Based on F. Baskett and J.L. Hennessy, Microprocessors: from desktops to supercomputers, Science 261:864–871, 13 August 1993.

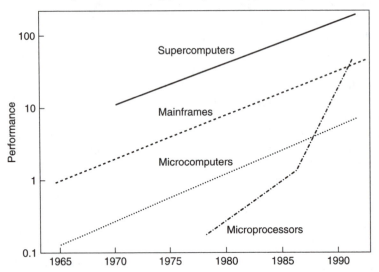

CPU Performance Trends
(millions of floating point operations per dollar)

FIGURE 6.8   CPU performance in floating point operations per dollar, plotted on a semi-log graph. Performance has continued to increase exponentially across a variety of technologies.
Based on B. Buzbee, Workstation clusters rise and shine, Science 261:852–853, 13 August 1993

somewhat the cost trends for materials and energy as a whole; Figure 6.11 indicates that prices for commodity materials, including petroleum, have tended to rise over the past 25 years, albeit with significant variability. Such data should, of course, be treated with considerable caution; substitution effects on all scales, complex patterns of technological evolution, and a thicket of direct and indirect subsidies complicate any analysis. Even if fuel and non-fuel material prices are assumed to remain stable, however, the cost differential favoring substitution of information and intellectual capital for other inputs should continue to grow because of the steep continuing decline in costs of the former.

There are a number of ad hoc examples, which as yet do not rise to proof of principle, illustrating how this fundamental evolution of the Industrial Revolution might be manifested. The substitution of telecommuting and satellite offices for lengthy employee commutes ("teleworking") is one; this trend has been encouraged in some countries by environmental laws, such as the Clean Air Act in the United States (subsequently amended to delete this provision). Such laws are important to overcome the barriers, which are largely cultural and not technological, to environmentally preferable working patterns. Another example is the use of information management systems, such as, sensor systems linked to computerized engine controls, in the modern automobile. This enables the product to produce its desired output, personal transportation, far more efficiently and environmentally preferably than only a decade ago. Some firms, such as AT&T, Microsoft, and IBM, are now offering software upgrades

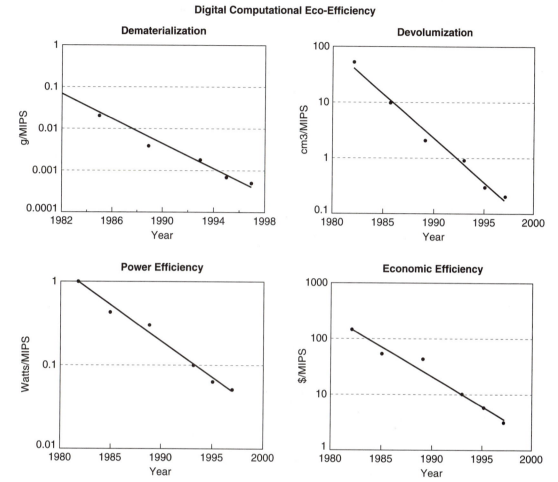

FIGURE 6.9 Digital computation technologies have continued to become more power efficient while using less material and becoming smaller and less expensive.

Source: David Lando, Bell Laboratories, Lucent Technologies

over telephone lines or the Internet rather than sending out a floppy disk, an example of substituting use of the telecommunications infrastructure for a physical product, or, in other words, dematerializing information transfer. Many journals now accept, if they do not require, submission of manuscripts in electronic form, and associated data appendices are routinely posted to the Internet. Indeed, some journals are virtually all online now, and some publishers, such as Epsilon Press in Surrey, England, have imprints dedicated completely to electronic publishing (in this case, it's called Eco-innovations Publishing). Although anecdotal, these and numerous other examples add support to the hypothesis that an economy increasingly based on information and intellectual capital tends to be less materials and energy intensive.

**Growth of Information Communication Equipment
Stocks, Japan, 1983-1993**

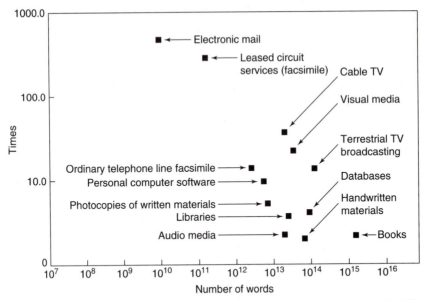

FIGURE 6.10 Growth of Information/Communication Equipment Stocks, Japan, 1983–1993
Source: Communications in Japan: 1995, Japan Ministry of Posts and Telecommunications, p16.

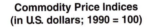

**Commodity Price Indices
(in U.S. dollars; 1990 = 100)**

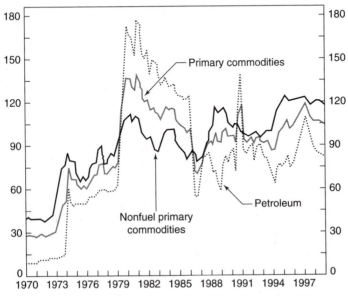

FIGURE 6.11 Costs of commodity materials. While costs per unit information processing and transmission capacity are falling rapidly, commodity and energy prices have remained roughly steady, or increasing slightly, over time.
Source: from World Economic Outlook: May 1997, International Monetary Fund, p.15.

6.8   GREEN CONSTRUCTION

Not surprisingly, the largest single flow of materials in the global economy is associated with construction materials. While in general not highly toxic, the sheer volume of material creates a number of environmental issues. For example, 20 to 30 percent by volume of many municipal landfills is construction waste and demolition debris. As in other sectors, the need to address these issues as part of normal planning, design, construction, operation, and end-of-life activities is increasingly being recognized, but the necessary tools and information are still primitive. It is also useful to differentiate between two major types of construction activities—facilities and infrastructure—although there is some overlap (e.g., port facilities or shopping malls). The former include buildings of all kinds—factories, schools, houses, fast food restaurants, banks—and raise common issues such as energy efficient fixtures and design, environmentally efficient operation, and design for upgrade, refurbishment, and reuse. The infrastructure category includes such projects as roads, gas pipelines, electricity distribution systems, water distribution systems, and the like, which in many cases may have significant secondary environmental impacts. A road into the jungle, for example, has the direct impacts associated with construction, but as infrastructure it also supports development along its route that would otherwise not have occurred.

Among the relevant differences between the two is the fact that an individual facility will often be placed where infrastructure dictates (factories near commercial transportation; stores near easy public access; built environments along major transportation corridors). Thus, placement of infrastructure may have an important impact on population dispersion patterns, which, in turn, may have significant implications for consumption of other goods and services. For example, the excellent road infrastructure of the United States combined with its expansive geography has led to a built environment where extensive automobile use is a virtual necessity. This has led some cities, such as Seattle in its recent draft Comprehensive Plan, to attempt to reduce reliance on personal transportation by basing the plan in large part on transportation infrastructure considerations. Fear of such patterns in Europe is a strong incentive for programs intended to create "urban villages" and maintain high quality of life in downtown areas, and has led to such proposals as banning suburban shopping malls in England. Such linked effects can only be addressed efficiently when policies are based on industrial ecology assessments that look at the system as an integrated whole.

Consider, for example, an industrial ecology approach to assessment of a proposed facility. The most fundamental need is to recognize the necessity of taking a life-cycle approach to built objects, just as with any other product. For this purpose, an Environmentally Responsible Facility (ERF) Matrix has been created (see Figure 15.4), with five life-cycle stages, including facility refurbishment, transfer, and closure at end-of-life, and five principle environmental concerns. For each cell in this five-by-five matrix, a rating from 0 (highest environmental impact, a very negative evaluation) to 4 (lowest impact, a very positive evaluation) is entered. The matrix-element values can then be summed, providing a figure of merit for the facility that can be compared to the maximum facility rating of 100. A set of checklists, keyed to each cell of the matrix and varying somewhat depending on the type of facility under consideration, supports the assessment. This is not the only such system; other building assessment methodologies include the Building Research Establishment Environmental

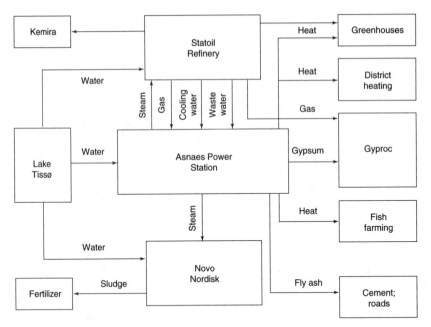

FIGURE 6.12   The industrial ecosystem (eco-park) at Kalundborg, Denmark.

Assessment Method, created in the United Kingdom with the support of major developers, and the Building Environmental Performance Assessment Criteria developed at the School of Architecture at the University of British Columbia in Canada.

Facility design and siting and infrastructure functions are combined in the concept of "industrial eco-parks," of which Kalundborg, in Denmark, is the most well-known. Figure 6.12 shows this geographically co-located system and some of the flows among the industrial nodes; similar efforts are being planned in other European countries and the United States. While such structures are desirable, their formation and dynamics are at present not well understood. In particular, it is not clear when the implicit assumption underlying such eco-parks—that geographical co-location is the most important factor in reducing environmental impact of such manufacturing systems and their concomitant residual flows—is valid. Ironically, their development is significantly impeded in some countries by existing environmental laws that impose disproportionate and burdensome regulation on residual flows, thus discouraging sharing and reuse of material.

6.9 ENERGY

The increasing availability of energy over the centuries has perhaps been the most critical contributor to an increasing quality of life for much of the world's population. Certainly, every modern economy depends on a secure, reliable energy infrastructure.

Fossil Fuel[1] Subsidies

	Subsidy rates in percent		Subsidy in million 1995 U.S. dollars		as Percent of GDP
	1991	1996	1991	1996	1996
Russia	45	31	28,797	9,427	1.5
Eastern Europe	42	23	13,120	5,838	3.19
China	42	20	24,545	10,297	2.42
India	25	19	4,250	2,663	1.06
Mexico	28	16	4,807	2,271	0.66
Brazil	23	0	1,951	11	0.0
OECD			12,453	9,890	0.05

[1]Includes petroleum products, natural gas, and coal

FIGURE 6.13   Fossil fuel subsidies, selected countries, 1991 and 1996. Although such subsidies are trending down, they still represent a large percent of GDP for many countries, and are environmentally inefficient as well.

Source: Based on World Bank, Expanding the Measure of Wealth: Indicators of Environmentally Sustainable Development (Discussion Draft), 1997

Energy is similar to built infrastructure in the sense that its form and availability determine many secondary characteristics of an economy and national population dispersion patterns; it is similar to agriculture in that it is a heavily subsidized and politicized economic sector. The rate of energy subsidies is particularly noteworthy in fossil fuels (Figure 6.13), which of course contributes to over-consumption of that resource as compared to others.

Combustion of fossil fuels (petroleum, coal, and natural gas) will be the predominant source of global energy production now and for the foreseeable future. Henry Linden at the Illinois Institute of Technology provides a snapshot of global energy consumption based on 1990-1991 data: world energy production and consumption are about 350 quadrillion BTUs, or "quads," annually, of which 140 quads is derived from liquid hydrocarbons; 75 quads from natural gas; 95 quads from coal; 20 quads from nuclear energy; and 25 quads from hydropower and other renewables. Nuclear energy is increasingly disfavored for economic and political reasons, and there are economic and technical reasons militating against rapid substitution of renewables for fossil fuels to any significant extent.

The primary driver for increasing energy demand is economic growth (Figure 6.14), which most projections show will continue, even accelerate, as Asia and Latin America develop. Concomitantly, most experts expect that reliance on fossil fuels will continue into the foreseeable future (Figure 6.15). Indeed, although there are significant allocation issues, and shortages resulting from demand and supply perturbations are always possible, it does not appear that an immediate absolute shortage of fossil fuel resources exists. Complacency is, however, misplaced: Conflict over strategic energy producing areas such as the Middle East and the Caspian Sea, and potential for substantial economic disruption because of competition for available energy products, especially petroleum, is not only possible, but likely.

Fossil fuel use does raise a number of environmental concerns, however. The major traditional concern is greenhouse forcing, primarily as a result of emissions of

World Energy, GDP, and Population Trends, 1970-2015

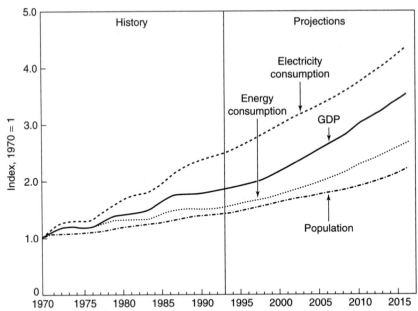

FIGURE 6.14   World energy, GDP, and population trends, 1970–2015.
Source: Energy Information Agency, International Outlook, 1996, p.5

World Energy Consumption by Fuel Type, 1970-2015

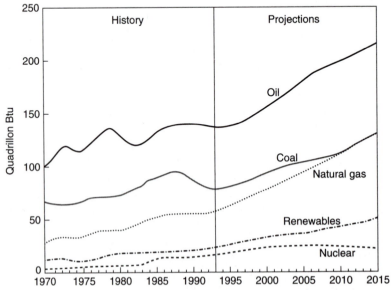

FIGURE 6.15   Reliance on fossil fuels is expected by most experts to continue, which places a high priority on efforts to reduce the attendant environmental impacts.
Source: Energy Information Agency, International Outlook, 1996

carbon dioxide and methane. A simple equation illustrates the relationships among population levels, energy use, and carbon emissions (usually in the form of carbon dioxide and methane, although other hydrocarbons may be involved):

Carbon emissions = population $\times$ GDP/capita $\times$ joules/unit GDP $\times$ carbon emissions/joule

In other words, total carbon emissions are a function of population levels, the standard of living, the energy efficiency of the economy, and the carbon intensity of energy production. As with the analogous equation for environmental impact presented in Chapter 3, the first two terms of this equation are trending upward, which means that the third and fourth, basically technology terms, must compensate not only for them, but for absolute levels of carbon emission that most scientists believe are already too high. Obviously, there are a number of important qualifications: Most importantly, per capita emissions depend heavily on a number of variables, including economic status, energy and transportation infrastructure, residential and employment patterns, climate, and consumption patterns (e.g., purchase of sports-utility vehicles in preference to sedans).

Currently, carbon emission levels are substantial. In 1991, for example, global use of 10.5 billion short tons of fossil fuel generated 26 billion short tons of carbon dioxide, or about 7.1 billion short tons of carbon, from the combustion of about 24 billion barrels of oil, 75 trillion cubic feet of natural gas, and 5.3 billion short tons of coal. Methane is a strong greenhouse gas with a global warming potential about 11 times that of carbon dioxide: about 20 of all global methane emissions, or about 100 million metric tons in 1990, are thought to be attributable to fossil fuel combustion.

Carbon emissions reflect economics and embedded plant design, however, rather than technological feasibility. There appear to be no fundamental scientific or technological reasons why carbon sequestration systems which strip carbon compounds form emission streams, then inject the liquified carbon into deep ocean or geologic sinks for storage, cannot be installed on centralized carbon fuel power plants (see Figure 8.3). If combined with a hydrogen energy system for mobile uses, it is even possible that fossil fuel could continue to be used over the medium term (decades to several centuries), thereby providing time for the development and deployment of environmentally preferable energy systems without contributing significantly to global climate change forcing.

Carbon emissions are not the only concern, however; other problematic emissions include sulfur and nitrogen oxides, which contribute to acid precipitation; and the emission and dispersion into the environment of heavy metals, particularly mercury, and radionuclides, primarily as a result of coal combustion.

These considerations need to be evaluated in light of energy use patterns at different stages of economic development, which exhibit two strong trends:

1. More developed economies are more energy intensive, although energy consumption can be highly price elastic, and production of services may be less energy intensive than a more capital intensive manufacturing economy.

2. As nations develop, energy consumption shifts from *primary* sources, such as direct consumption of oil, biomass, or coal, to the *secondary* source, electricity, as

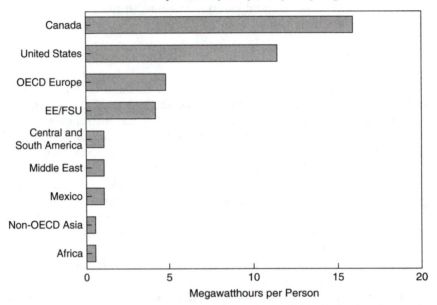

FIGURE 6.16   High per capita electricity consumption is associated with developed economies, as well as the physical characteristics of a country (large grids in the U.S. and Canada have high transmission losses). Source: Energy Information Administration, International Energy Outlook, 1996

demonstrated by per capita consumption figures (Figure 6.16). Reflecting older technology systems, the electrical production and distribution system has traditionally been heavily centralized in developed economies, making environmental controls imposed on point sources relatively easy. Currently, however, a number of technologies that can be easily implemented on a small scale have resulted in strong pressures to decentralize the electricity production functions. The environmental implications of this sectoral evolution are not yet clear, and may well depend on the regulatory structure within which it takes place.

REFERENCES

Bormann, B.T., M.H. Brookes, E.D. Ford, A.R. Kiester, C.D. Oliver, and J.F. Weigand. *A Broad, Strategic Framework for Sustainable-Ecosystem Management.* Prepared by Eastside Forest Health Panel for USDA Forest Service, May, 1993.

Canadian Standards Association. *A Sustainable Forest Management System: Guidance Document and Specification Document.* CAN/CSA-Z808-96 and CAN/CSA-Z809-96, 1996.

Crosson, P.R. "Is U.S. agriculture sustainable?" *Resources.* Fall 1994, pp. 10 et. seq.

Faeth, P. *Growing Green: Enhancing the Economic and Environmental Performance of U. S. Agriculture.* Washington, DC, World Resources Institute: 1995.

Falk, N., M. Chase, and S. Hampson. "Are our town centres sustainable?" *RSA Journal.* 1995, CXLIII(5461):35–55.

Fanney, A.H., K.M. Whitter, A.E. Travgott, and L.N. Simon, eds. *U. S. Green Building Conference–1994*. Washington, DC, National Institute of Standards and Technology: 1994.

Gertler, N. *Industrial Ecosystems: Developing Sustainable Industrial Structures*. Masters thesis, MIT, May 1995.

Graedel, T.E. and B.R. Allenby. *Industrial Ecology*. Upper Saddle River, NJ, Prentice-Hall: 1995.

Herzog, H.J. and E.M. Drake. "Carbon dioxide recovery and disposal from large energy systems." *Annual Review of Energy and Environment*. 21:145–66 (1996).

Hodges, C.A. "Mineral resources, environmental issues, and land use." *Science*. 268:1305–1312.

IEEE White Paper on Sustainable Development and Industrial Ecology, Technical Activities Board of the Institute of Electrical and Electronics Engineers, Inc., Washington, DC, 1995.

Klimisch, R.L. "Designing the modern automobile for recycling," in B.R. Allenby and D. J. Richards, eds. *The Greening of Industrial Ecosystems*. Washington, DC, National Academy Press: 1994, pp. 165–171.

Linden, H.R. "Energy and industrial ecology," in B.R. Allenby and D.J. Richards, eds. *The Greening of Industrial Ecosystems*. Washington, DC, National Academy Press: 1994, pp. 38–60.

Meikle, J.L. *American Plastic: A Cultural History*. New Brunswick, NJ, Rutgers University Press: 1995.

Public Technology, Inc. and U.S. Green Building Council. *Local Government Sustainable Buildings Guidebook: Environmentally Responsible Building Design and Management*. Washington, DC: 1993.

Rothschild, B.J. "How bountiful are ocean fisheries?" *Consequences*. 2(1):14–24, 1996.

St. John, A., editor. *The Sourcebook for Sustainable Design: A Guide to Environmentally Responsible Building Materials and Processes*. Boston, Boston Society of Architects: 1992.

U.S. Energy Information Administration. *International Energy Outlook, 1996*. Washington, DC, US Government Printing Office: 1996.

EXERCISES

1. Industrial ecologists believe that some industrial sectors will grow, while others must shrink, as society moves towards sustainability. What sectors do you think will grow and shrink, and why?

2. If you were a private firm in one of the shrinking sectors identified in the above question, what could you do to minimize the impacts on your operations in the short term? In the long term? Are your proposed actions good for society as well? Why or why not? If they are not, how can you redefine your business so that your interests and society's interests are the same?

3. Policy and analysis tends to focus on specific sectors, a division frequently enshrined in government structures, where different ministries are established to deal with individual important sectors such as energy, agriculture, and transportation. Consider yourself a planner in an environmentally sensitive Northern European country, with the responsibility of establishing policies that will encourage the evolution of sustainable cities. What sectors must you integrate into your plans? Using your government as a model, determine how a viable multi-sectoral policy could be developed and implemented given the fragmented nature of the governmental structure.

CHAPTER 7

Applications to Practice: Techniques and Tools

7.1 THE ROLE OF TECHNIQUES AND TOOLS

Seeking to build a sustainable global economy requires more than simply good intentions and desire, or repeating the mantra that environmental issues are now strategic rather than overhead. It requires that a comprehensive set of tools and techniques be developed based on the principles of industrial ecology and the scientific and technological insights the study of that field generates. These techniques and tools form the mechanism by which the findings of industrial ecology and environmental and physical science, and the political and cultural acceptance of sustainability as a legitimate goal, can be translated into the institutions and practices that form the basis of our economy. Most importantly, a set of linked indicators and metrics at different systems levels appear to be the most feasible, if not the only, mechanism by which policy initiatives can be robustly integrated to support the achievement of sustainability.

The extent to which such techniques and tools are being explored is frequently not realized by those not in the field of industrial ecology or its application in various sectors. A number of qualitative, even semi-quantitative, tools have been developed, for example, in the areas of life-cycle assessment, or LCA, and Design for Environment, or DFE. On the other hand, it is apparent that virtually all tools and techniques currently in use are at a very primitive stage, and the results of such analyses, even where relatively comprehensive, must be viewed skeptically. Existing data are frequently sparse and of dubious quality; in many cases, the most critical data do not exist. Even where adequate data does exist, methodologies to convert them into understanding and knowledge are inadequate. In other cases, large amounts of trivial or relatively irrelevant data overwhelm the analytical process, producing voluminous results but little knowledge.

The issue of incompatible value systems, which leads to widely differing values being placed on different options by different people, is equally as important but more difficult to resolve. Such dimensions lie in the province of politics and ethics, and not science and technology, although they should be informed by them. Thus, for example, an economist might put no value on the existence of a species if no one is willing to pay money to preserve it, while, to an environmentalist, a species may be inherently above valuation and its destruction ethically unacceptable. Similarly, many people may have trouble accepting the suggestion that a human life is, in practice, worth a finite sum of money, and, more difficult yet, that the sums differ depending on culture and socio-economic status.

This chapter introduces a number of generic techniques and tools that are currently being developed or used. It must be remembered, however, that this process is in its infancy. Accordingly, these approaches should be taken as illustrative. It is quite likely that, as the field of industrial ecology advances, the complexity and diversity of tools will expand considerably; and that some of the techniques discussed, such as genetic engineering, will be applied in ways, and produce results, that are beyond our ability to forecast currently.

7.2 SUSTAINABILITY INDICATORS, METRICS, AND SENSORS

Determining what dimensions of a changing complex system are meaningful depends on several factors, most critically the reasons the system matters to different stakeholders and the ability to understand the system in terms of those reasons. In this case, the fundamental reason for the study of economic and underlying natural systems and their interactions through the field of industrial ecology is to understand what a sustainable system might look like. Accepting this as the goal, the question then becomes how to understand an extraordinarily complex system and, equally important, how to develop policies that can respond to this understanding to move closer to the goal of sustainability. The means by which this can be accomplished is through the use of *indicators*, *metrics*, and *sensors*. Without such guides, it is almost impossible to generate meaningful policies; thus, a fundamental initial policy goal should be to establish a set of indicators and metrics by which one can evaluate progress toward sustainability goals, and develop and modify subsequent policies.

7.2.1 Integrated Indicator Systems

The initial difficulty in discussing indicator systems is semantic: a number of different terms—"themes," "goals," "indicators," "indices," "metrics"—are used somewhat interchangeably to refer to different levels of data aggregation. Moreover, the critical difference between normative, desired states—"this is where we ought to be"—and objective assessments of state—"this is where we are, or will be if present trends continue"—can be obscured by this proliferation of terms. The situation is further confused because some terms, such as "indicator," are technical terms used in environmental science fields such as ecosystem toxicology. An "indicator species," for example, is one that indicates, through its presence or otherwise, something about the state of the environment within which it is found.

Nonetheless, the conceptual structure of a comprehensive indicator system is not complex. It begins with the recognition that the only way a rational policy structure parallel to the industrial ecology framework introduced in Chapter 2 can be created is to generate a set of linked goals, metrics, and sensor systems, and create databases that support each level of the hierarchy (Figure 7.1). Policies that support sustainability

Linking System Levels

Indicator network			Industrial ecology system levels	Policy levels
Goals	**Measures of success**	**Implementation**		
Sustainability indicators and indices	↔ Sustainability metrics	↔ Sensor systems and data gathering at appropriate scale	Sustainable development	High level sustainability policies
↕	↕	↕	↕	↕
Industrial ecology indicators and indices	↔ Industrial ecology metrics	↔ Sensor systems and data gathering at appropriate scale	Industrial ecology	Industrial ecology policies
↕	↕	↕	↕	↕
Industrial ecology infrastructure indicators and indices	↔ Industrial ecology infrastructure metrics	↔ Sensor systems and data gathering at appropriate scale	Industrial ecology infrastructure	Industrial ecology implementation policies
↕	↕	↕	↕	↕
Sector, technology and activity-specific indicators and indices	↔ Sector, technology and activity-specific metrics	↔ Sensor systems and data gathering at appropriate scale, including performance data	Application to practice	Targeted policies (e.g., sector, technology or issue-specific)

FIGURE 7.1   Creating networks of metrics, indicators, and sensors at different relevant levels of natural systems is an important, perhaps required, step towards real environmental improvements. It is also necessary to support any rational policy regime.

directly, for example, are not likely to be much help in designing a manufacturing process to clean printed wiring boards in electronics. Nonetheless, if a system of metrics exists that links aspects of that process (for example, measured lower energy consumption) through intermediate levels (reductions in emissions of carbon dioxide from electricity generation) to sustainability goals (reduced global climate change forcing), one can at least ensure that policy initiatives taken at different levels integrate into a coherent policy structure. In the absence of such a system, policy alignment across very different and complex systems levels will be difficult, if not impossible, to achieve.

As one moves up the hierarchy from specific, short-term applications to broad, social goals of sustainability, vast amounts of data are condensed into increasingly fewer indicators (Figure 7.2). For example, information on carbon dioxide emissions from individual processes and point sources becomes integrated into single figures on national emissions. Moreover, the figures become increasingly normative: The emissions of carbon dioxide from individual processes is objective, factual data, while the goal of reducing carbon dioxide emissions, which is the motivator for gathering such data in the first place, is relatively normative. Additionally, the primary audiences tend to differ: The public understands highly aggregated index numbers or indicators, but has difficulty dealing with issues of data quality and quantity, uncertainty, and relative

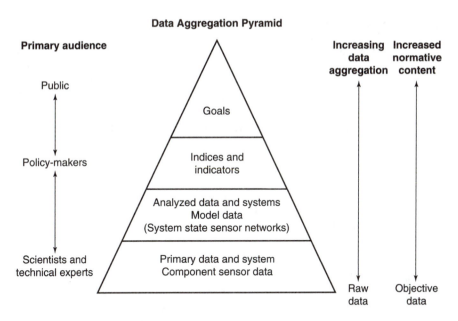

FIGURE 7.2 Aggregating data is both technically difficult and, because of the increasing normative content of high level goals and metrics, a challenge to policy formulation as well.

risk. Scientists and technical experts, on the other hand, frequently want the information that the data aggregation process necessarily washes out.

7.2.2 Indicator System Development

Although it is still a nascent area of research, a few examples from the work that has already been done on indicator systems and metrics will clarify the previous discussion. The first real efforts to develop sustainability indicators began in the late 1980s, primarily by the Canadian and Dutch governments, as well as the OECD. In 1987, for example, the Dutch began work on such indicators as part of their effort to formulate the National Environmental Policy Plan of 1989, with a goal of making the country sustainable in a generation. Obviously, this required some metrics by which progress toward that admittedly vague goal could be measured. As discussed by Adriaanse, it was eventually determined that such indicators should meet four requirements:

1. They must be of high quality, that is, they must bear some relation to underlying causal relationships within the system being studied, and must aggregate as much information as possible into a meaningful composite measure, which should have intuitive appeal and be easy to understand.

2. They must accurately reflect a trend with an appropriate time scale (e.g., a metric for global climate change should not be calibrated, or vary, over a time scale of months) and should, where applicable, display both medium-term and long-term effects.

3. They must link to existing policy objectives and related activities and, critically, ultimately link to the achievement of sustainability (however, society defines that term). It is this latter feature that standard environmental indicators lack.

4. They must be clear and understandable by the public, both to engender support for the policies that are associated with them and to achieve, over time, the culture change that will be a prerequisite to a sustainable global economy.

The "themes" for which indicators were developed in the Netherlands include the following:

- Climate change
- Acidification of the environment
- Eutrophication (a particular problem in the Netherlands)
- Dispersion of toxics
- Solid waste accumulation
- Disturbance of local environments
- Dehydration of soils
- Resource depletion

Each theme then had a quantitative indicator and targets developed for it, which might have a number of underlying constituents that in turn might be based on aggregated data. Thus, for example, the climate change theme has an overall "pressure" indicator, which in turn is made up of components based on carbon dioxide emissions, methane emissions, nitrous oxide emissions, and CFC and halon production. Thus, for example, Figure 7.3 shows the Dutch calculation for their climate change theme, and Figures 7.4 through 7.7 show the summed calculations for the different classes of major greenhouse gases—nitrous oxide, chlorofluorocarbons, carbon dioxide, and methane— as well as targets for reductions. The result is a clear and concise statement of progress towards the defined social goal.

Developing such indicators requires an understanding of not only scientific data, but also the context within which such indicators work, including economic, technical, policy, and cultural dimensions. Hammond et al. of The World Resources Institute express this formulation as the Pressure-State-Response Framework (Figure 7.8), and, drawing on work done at the OECD and UNEP, have produced a broader matrix of sustainability indicators, presented in Table 7.1. While other indicators can and have been suggested, it remains important in all cases to concentrate on those few that meet the criteria suggested above, and not to generate so much information that the requisite aggregation necessary to communicate easily with the lay public is lost.

The appropriateness of different indicators will vary along a number of dimensions, including local and regional differences in physical characteristics, demographic patterns, state of economic development, trade patterns, and others. Moreover, the development of such indicators is still in its infancy, and there are numerous issues that remain to be worked out. Most fundamentally, because society cannot yet define a state of sustainability rigorously, it is impossible to develop a set of rigorous metrics by which distance from that endpoint can be determined. It is clearly possible, however, to

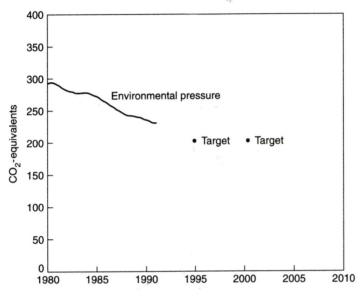

FIGURE 7.3   Climate change theme
Based on A. Adriaanse, *Environmental Policy Performance Indicators*, Sdu Uitgeverij
Koninginnegracht, 1993.

FIGURE 7.4   Nitrogen oxides emissions
Based on A. Adriaanse, *Environmental Policy Performance Indicators*, Sdu Uitgeverij
Koninginnegracht, 1993.

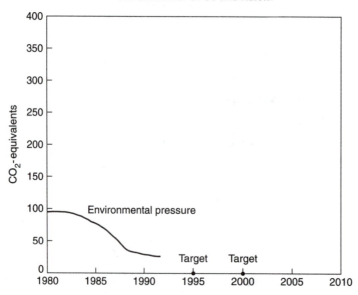

FIGURE 7.5 Production of CFCs and halons data
Based on A. Adriaanse, *Environmental Policy Performance Indicators*, Sdu Uitgeverij Koninginnegracht, 1993.

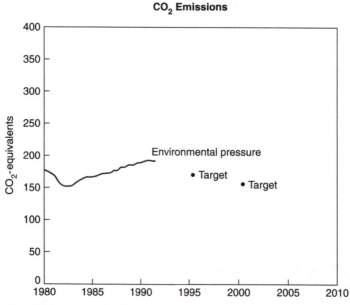

FIGURE 7.6 Carbon dioxide emissions
Based on A. Adriaanse, *Environmental Policy Performance Indicators*, Sdu Uitgeverij Koninginnegracht, 1993.

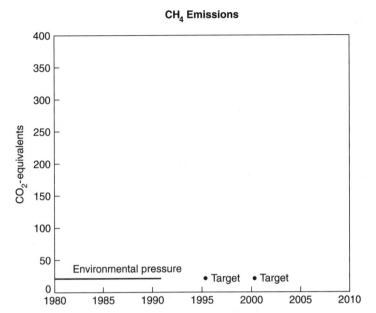

FIGURE 7.7   Methane emissions
Based on A. Adriaanse, *Environmental Policy Performance Indicators*, Sdu Uitgeverij
Koninginnegracht, 1993.

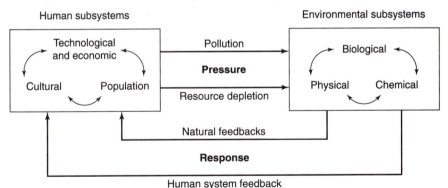

FIGURE 7.8   From a systems perspective, the desire is to strengthen feedback loops between human and
natural systems to attempt to achieve stability of desirable states.
Modified from a model by Hammond, Adriaanse, Rodenburg, Bryant and Woodward,
Environmental Indicators, World Resources Institute, 1995.

determine the types of metrics that can measure progress toward that general goal, and
to quantitatively measure improvements in such metrics, even if the quantitative validity of specific targets cannot be absolutely ascertained. As with anything in its early

TABLE 7.1 Matrix of Environmental Indicators

Issues	Pressure	State	Response
Climate Change	Greenhouse gas emissions	Concentrations	Energy intensity
Ozone Depletion	Halocarbon emissions	Chlorine concentrations in stratosphere; O_3 column	Montreal Protocol; CFC recovery
Eutrophication	Nitrogen, Phosphorous emissions	Nitrogen, Phosphorous concentrations; Biological Oxygen Demand	Water treatment investments and costs
Acidification	SO_X, NO_X, NH_3 emissions	Deposition; concentrations	Investments; control agreements
Toxic Contamination	Heavy metal, persistent organic compounds (POC) emissions	POC, heavy metal concentrations	Recovery of hazardous waste; control and investments/costs
Urban Quality of Life	Volatile organic compounds (VOC), NO_X, SO_X emissions	VOC, NO_X, SO_X concentrations	Expenditures; transportation policy
Biodiversity	Land conversion; land fragmentation	Species abundance	Protected areas
Waste	Waste generation by sectors and communities	Soil/groundwater quality	Material collection rate; recycling investments and costs
Water Resources	Demand and use intensity by sector and communities	Demand/supply ratio by sector; water quality	Expenditures; water pricing; savings policy
Forest Resources	Use intensity	Area of degraded forest	Protected area of forest; prevalence of sustainable logging practices
Fish Resources	Fish catches	Sustainable stocks	Quotas and economic rationalization
Soil Degradation	Land use Changes	Top soil loss; soil degradation	Rehabilitation/ protection
Oceans/Coastal Zones	Emissions; oil spills; depositions; eutrophication	Water quality; biological impacts	Coastal zone management; ocean protection
Environmental Index	Pressure index	State index	Response index

Based on Hammond, Adriaanse, Rodenburg, Bryant and Woodward, *Environmental Indicators*, World Resources Institute, 1995.

stages of development, it is important that the desire to achieve the eventual best not be allowed to disrupt the implementation of the good.

7.3 ECONOMIC REFORM

7.3.1 Externalities

It is axiomatic among economists of virtually all stripes that much that is dysfunctional in the existing relationship between economic activity and the local, regional, and global environment could be mitigated if environmental impacts that are not currently captured in the costs of services and products, called "externalities", were to be priced and captured. Some have even argued that doing so—"internalizing the externalities" —would in itself be sufficient to achieve sustainability. Four major criticisms have been made against this theoretical position:

1. Significant price adjustments, such as substantially raising the price of energy to reflect externalities through a so-called "carbon tax," have proven to be politically impossible. Minor adjustments have been made—the United States imposed a tax on CFCs as part of the phase-out process, and some Northern European countries have imposed a limited carbon/energy tax—but no large scale, direct incorporation of environmental externalities into the economic system has occurred in any nation.

2. The data and methodologies by which externalities can be adequately quantified preparatory to incorporating them into the price structure are usually weak, when they are not altogether lacking.

3. Quantifying environmental impacts requires that one establish an accepted underlying values prioritization—how much money a species is worth, or, for that matter, a human life. In fact, existing systems that attempt to perform such evaluations pursuant to ranking the environmental performance of materials, for example, usually rely on a "willingness-to-pay" methodology. While obviously necessary if one is to improve the environmental performance of the economy, such quantifications need to be regarded as contingent, and are unacceptable to some people.

4. If the natural systems subject to anthropogenic forcing—the global climate system, for example—are deterministically chaotic, they are inherently unpredictable at the granularity at which costs will be imposed. Thus, costs cannot be known until they have already been borne, at which point, of course, it is too late to modify the behavior leading to them in order to avoid them. In such a case, externalities simply do not occur until after it is too late to internalize them. For example, if a consequence of global climate change is that California wineries are wiped out, but this cannot be determined until it has, in fact, occurred, the price structure of carbon dioxide emissions (e.g., a carbon tax) cannot be modified beforehand to avoid that cost.

Nonetheless, it is certainly true that to the extent externalities can be identified, quantified, and driven into the costs of the economic activity that produce them, environmental and economic efficiency are both served (although there may well be significant costs associated with the transition). In fact, much environmental regulation can be seen in economic terms as simply a grossly inefficient effort to compensate for failure of prices to capture such externalities. Moreover, in most cases, enough information is clearly known to begin moving costs in the right direction, even if the absolutely accurate value cannot be ascertained. For example, it is obvious by any measure that energy prices in many countries are too low given the environmental externalities associated with energy production and consumption, and that process and product technologies are therefore skewed toward environmentally inefficient designs as a result.

There are three general mechanisms, therefore, that can help achieve greater environmental and economic efficiency. The first is capturing externalities directly in the price structure through tax systems, subsidies, or other governmental fiscal policy tools. This is an area that has been extensively commented on in other texts, and need not be considered further here.

The second mechanism has been identified in both the sustainable agriculture and the integrated materials management discussions: elimination of subsidies. In most cases, subsidies are established to help short-term development of sectors deemed to be critical to economic growth or national security: food (agriculture) and energy are the most common, but other sectors, particularly the extractive and transportation sectors, are frequently favored. The effect of a subsidy, of course, is to lower the price of the targeted commodity, product, or service below that which the market would ordinarily establish. Because in many cases these sectors already have significant environmental externalities associated with them, lowering the costs of production even more simply results in greater environmental impacts. Both economic and environmental efficiency suffer, making such subsidies doubly indefensible.

7.3.2 Green Accounting Systems

The third mechanism is known as *green accounting systems*. Contrary to some popular usage, these are not accounting systems that incorporate externalities directly; that requires defining and internalizing costs that have not yet been monetized and incorporated into the economic system, and is properly considered as a social issue, not an accounting issue. Strictly speaking, accounting only deals with economic costs. Rather, green accounting systems identify costs that are in fact already accrued, but for methodological reasons are not yet recognized as costs, and then make them apparent to policymakers and managers. This occurs at a number of levels.

At the level of the individual firm, most environmental costs today are simply lumped together in a residual accounting category known as "overhead." They are therefore not focused on by managers, and changes in operations that would improve both economic and environmental efficiency are not identified, nor are financial or managerial incentives for their implementation created. What is required is management accounting systems that break out existing environmental costs, such as liability and compliance costs associated with specific waste streams, assign them to the causative activity, and thus permit their rational management.

In practice, this is a difficult task. The physical data on the contributions of different processes and products to a residue stream frequently don't exist, and may be technically hard to obtain. Managers tend to resist any increase in the number of elements of the business process for which they will be held responsible (they like having things in overhead accounts). Also, assigning potential future liabilities back to existing accounts might be resisted for fear it would increase legal liability: It might be argued that a company that did so was thereby admitting its planned behavior was inappropriate or illegal. Nonetheless, it is apparent that development of appropriate managerial accounting systems is critical to completing a necessary feedback loop for environmentally appropriate behavior by corporations, and that wise public policy would encourage such an evolution.

At the national level, accounting systems are generally based on the international standard for national accounts, the United Nations System of National Accounts

(SNA). The SNA recognize land, mineral, and timber resources as assets in a country's capital stock, but do not recognize them in the income and product accounts. Accordingly, as resources are used and thereby generate a flow of income, they are not depreciated in the income and product accounts. Satellite accounts to the SNA to allow national environmental accounting are being developed, but they are currently separate from the core SNA accounts. Figure 7.9 from the SNA outlines the physical accounts that would underlie such a national accounts system. In some countries, particularly those that rely on resource depletion to generate current income, this disparate treatment of resources creates significantly skewed assessments of the real economic costs of policy decisions.

At the international level, no accounting system capable of supporting the evaluation of international costs and benefits arising from the mitigation of regional and global environmental perturbations exists. Given the political, cultural, and economic

Physical Accounts as Component of Natural Resource Accounts
Algorithm: 1 + 2 − 3 ± 4 = 5

			Biological assets		Subsoil assets (proved reserves)	Water		Air		Land (including ecosystems)		
										Cultivated		Uncultivated (area)
			Produced	Wild		Quantities	Qualities (constituents)	Quantities	Qualities (constituents)	Soil	Area	
			1	2	3	4	5	6	7	8	9	10
1	1	Opening stocks	X	X	X	X	X		X	X	X	X
	+2	*Increase*										
2	2.1	Gross natural increase	X	X	X	X				X	X	X
3	2.2	Discovery of resources		X	X	X						
4	2.3	Area increase by economic influence									X	X
	−3	*Decrease*	X	X	X	X				X	X	X
5	3.1	Decrease due to natural causes										
6	3.2	Depletion due to economic causes	X	X	X	X				X		
7	3.3	Area decrease by economic influence									X	X
	+/−4	*Adjustments*										
8	4.1	Technical improvements		X	X	X						
9	4.2	Changes in prices, costs		X	X	X						
10	4.3	Improved estimation methods	X	X	X	X				X	X	X
11	=5	Closing stocks	X	X	X	X	X		X	X	X	X

FIGURE 7.9   Physical accounts as a component of natural resource accounts. Such methods are not yet in routine use.
Based on UN Handbook of National Accounting: Integrated Environmental and Economic Accounting, 1993

differences among nations, developing such an accounting system is a daunting task. Yet it could be well worthwhile if it could support international mitigation efforts targeted at specific national activities, while being of clear economic benefit to other nations. For example, it would be extremely useful to have an accounting system that would reveal the conditions under which it might be in the economic self-interest of Europe and the United States to fund clean coal technologies in India and China. Similarly, Japan has a clear interest in understanding the costs and benefits of funding technologies in China that reduce that nation's substantial contribution to acid rain in Japan.

7.4   DESIGN FOR ENVIRONMENT AND LIFE-CYCLE ASSESSMENT TOOLS

There are a growing number of tools that can be used to perform studies of the environmental impacts of processes, products, and materials, usually over their life cycle. These range from the highly detailed *LCA* (*life-cycle assessment* or analysis) methodologies developed by the Society of Environmental Toxicology and Chemistry to DFE methodologies developed by companies such as AT&T and Volvo to provide *abbreviated life-cycle assessments* (*ACLA*), which are useful in the time and resource constraints imposed by a commercial environment. Some methodologies, such as that developed by SETAC, are intended to capture environmental effects only and are most useful when applied to a single material in a specific use. Some, such as the matrix system developed by Allenby, attempt to integrate social, commercial, and environmental impacts in the analysis. Others, such as the AT&T matrix system, attempt to capture major environmental impacts in a concise fashion, and can be applied efficiently to complex manufactured products. Some, like the Swedish/Volvo Environmental Priority Strategies for Product Design (EPS) system, provide results in the form of an Environmental Load Unit for a particular design choice, making integration into computerized design systems easy. Finally, others, such as the VNCI (Vereniging van de Nederlandse Chemische Industrie) system, combine environmental and economic profiles to identify the most environmentally and economically efficient options.

Conceptually, each of these systems addresses in some way three basic components of any life-cycle assessment. First, they attempt with varying degrees of detail to identify the releases to the environment and other environmental implications (e.g., filling wetlands) that result from the option under consideration: the *inventory analysis*. Next, they attempt to relate this data to changes in the world external to the option: the *impact analysis*. Finally, they determine how improvements can be made: the *improvement analysis*. Sometimes this last stage only considers improvements to the option under investigation; other times the range of options that might be substituted are also included in the analysis. Obviously, the latter is preferable and more realistic in a highly competitive global economy, even if it frequently complicates the analysis.

The inventory analysis and impact analysis stages can be relatively rigorous, assuming sufficient data can be collected or developed. Linking the impact analysis with the improvement analysis, however, is where the difficult, non-quantitative issue of values arises. Improvement, after all, not only implies reductions in any one emission, but also trade-offs among environmental impacts, and between environmental impacts and other social goals such as economic efficiency and employment, so that a higher quality of life is created. It is thus far more difficult and contentious, and care must be taken not to assume that a merely technical manipulation of data can resolve issues arising at this stage.

Each of these DFE and LCA systems, which are described more fully in the references listed at the end of the chapter, has its strengths and weaknesses. Some, such as the SETAC system, are very complex and detailed, and thus require a long time (around six months in many cases) and substantial resources to complete. Others, such as the AT&T matrix systems, are concise and relatively easy to perform (frequently one or two days), but, being more qualitative and judgmental, require a trained industrial ecologist. All should be regarded at this point as experimental, and the natural tendency to rely primarily on any one approach should be rejected as premature.

It is, in fact, likely that many of the specific results attained in the DFE and LCA analyses performed to date are inaccurate, or at least incomplete (that is, they do not reflect the environmental impacts when the system is viewed as a whole) because critical data are weak or non-existent and methodologies are still quite primitive. Nonetheless, these methods are valuable in helping individuals, firms, and governments move up a very complex learning curve, and in changing their cultures so that the fundamental requirement of systems-based, life-cycle oriented thinking is encouraged. Moreover, use of these methodologies is frequently valuable in confounding conventional wisdom, thus demonstrating how profound the existing ignorance actually is. Thus, Allenby's study of lead solder use in electronics manufacture indicated that it might well be preferable to bismuth or indium-based alternatives, and a Dutch study showed that, in many cases, plastic and paper cups are preferable to porcelain alternatives. Both of these results are, of course, counterintuitive.

This is a period of almost baroque experimentation with LCA and DFE approaches, a process that, although untidy, is providing a great deal of knowledge and experience, and is also building a community of practitioners and scholars. Accordingly, the policy goal should be to encourage the use of such methodologies, while not discouraging experimentation and future evolution of these tools. It is thus encouraging that the International Standards Organization (ISO) is establishing a set of standards under the generic designation of ISO 14000 based on the life-cycle concept, as this should encourage the diffusion of this approach.

7.5   BIOLOGICAL ENGINEERING

Traditionally, the primary economic sectors concerned with biology in their production activities were the agricultural and forestry sectors, and, of course, the health-care sector. Two trends, both of which may be anticipated to accelerate in an environmentally constrained world, are changing the role of biological entities in the economy: (1) genetic engineering and (2) integration of biological and engineered systems.

The potential for genetic engineering to revolutionize economic production—or, for that matter, the extent to which it already has—is not yet fully recognized. What has changed is not the fact that genetic engineering is performed: Virtually all modern commercially important crops are the result of genetic engineering. Maize, for example, has been so changed over the centuries that modern strains cannot reproduce without active human intervention. What has changed is the speed and specificity with which genetic characteristics can be developed and transferred among organisms, the capability to shift xenogenic material among species, and the growing scientific underpinning of this process. Naturally, a primary focus has been on improving the performance of existing crops, especially regarding pest resistance and herbicide tolerance (so that strong applications of herbicide can be used to kill undesired species while not adversely affecting the crop). Such technologies may enable substantial reductions in the amount of energy, fertilizer, or pesticides required per unit product. More fundamental possibilities exist, however. For example, sheep and goats have been genetically engineered to produce drugs ("biofactories on the farm"); bacteria have been modified to produce plastics; and cotton has been genetically engineered to express polyester in its fibers.

As with any new technology, concerns have been raised about genetic engineering. Farmers fearing competition from more efficient cultivars have supported trade restrictions against so-called GMOs (Genetically Modified Organisms), and, in alliance with environmentalists, have had particular success in Europe. More fundamentally, ecologists are concerned that not enough may be known about interspecies genetic transfer mechanisms; some, for example, fear that wild weeds might pick up the genetically engineered pest resistance from engineered crop plants, thereby further threatening agricultural production. Others have expressed ethical concerns about humans manipulating genetic material at all, a concern that grows stronger the closer the species is to *homo sapiens*; an effect that can clearly be seen in the strong responses to the first successful cloning of a sheep from adult genetic material.

A second category of biological engineering worth noting is the inclusion of biological components in engineered systems. One obvious example is the use of engineered wetlands as waste treatment facilities, substituting for expensive sewage treatment plants. Another is the use of biological systems in the life-cycle management of metals. Unbeknownst to many people, bacteria have long been used commercially to extract metals such as gold from ore. Now, however, the use of certain plants—such as Alpine pennycress (*Thlaspi caerulescens*) for zinc and cadmium, certain *Brassica* species for selenium, lead, chromium, nickel, zinc, copper, and cadmium—to remove and concentrate metal residuals and other pollutants from contaminated soil is under investigation. The plants can then be harvested and dried, and the metal either recovered or segregated as hazardous waste (a decision that will revolve around regulatory burdens and the value of the captured metal). More broadly, it is likely that, over time, biomass plantations, possibly growing genetically altered species, will increasingly provide material for energy production as well as chemical feedstocks. The end result of this trend is suggested by a 1970s NASA study of space colonies, which foresaw a built environment where biological systems and engineered systems were fully integrated. Figure 7.10 provides a flow sheet example of a generalized closed life-support system from that study.

Biological systems have significant advantages over many engineered systems, especially in materials processing and production. They tend to substitute finesse for brute force; for example, using enzyme systems at room temperature to perform the same chemistry as high pressure, high temperature, energy-intensive commercial

NASA Closed Loop Life Support System
(T_1 through T_7 mark transformation points)

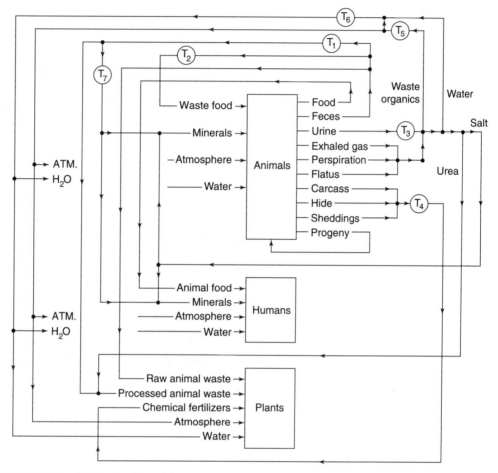

FIGURE 7.10   NASA closed loop life support system
Source: NASA, Space Resources and Space Settlements, NASA SP-428, 1979

processes that frequently involve toxic feedstocks, intermediaries, or residual streams. Food, fiber, materials, and products can be produced with (in most cases) far less environmental impact, an important consideration given the inexorable increase in human population and desire for material consumption. For these reasons, it is likely that the trends favoring development of GMOs and the substitution of biological for traditionally engineered systems will accelerate.

7.6   COMPREHENSIVE RISK ASSESSMENT METHODOLOGIES

The subject of risk assessments and cost/benefit analyses is covered in more detail in Chapter 10. It is important to recognize here, however, that without the development of Comprehensive Risk Assessments (CRAs), which not only integrate evaluations among environmental risks, but also between environmental and other forms of risk (economic, employment, or cultural, for example), it will be very difficult to achieve significant progress toward sustainability. This interdependence of factors is captured in the concept of the "triple bottom line"—social, economic, and environmental efficiency and returns—which is becoming increasingly popular, especially in Europe. To evaluate options that involve trade-offs of many different kinds, something like a CRA will eventually be required. As with human perturbations of such fundamental natural systems as the carbon cycle, such trade-offs are, in fact, being made every day; they are just being made unconsciously and poorly. Accordingly, development of such tools should have a high priority in any national policy seeking to approach sustainability.

REFERENCES

Adriaanse, A. *Environmental Policy Performance Indicators: A Study on the Development of Indicators for Environmental Policy in the Netherlands*. Sdu Uitgeverij Koninginnegracht, 1993.

Allenby, B.R. Design for Environment: Implementing Industrial Ecology. Doctoral Dissertation, Rutgers University: 1992.

Daly, H.E., and J.B. Cobb, Jr. *For the Common Good*. Boston, Beacon Press: 1989.

Graedel, T.E. and B.R. Allenby. *Industrial Ecology*. Upper Saddle River, NJ, Prentice-Hall: 1995.

Hammond, A., A. Adriaanse, E. Rodenburg, D. Bryant, and R, Woodward. *Environmental Indicators: A Systematic Approach to Measuring and Reporting on Environmental Policy Performance in the Context of Sustainable Development*. Washington, DC, World Resources Institute: 1995.

Keoleian, G.A. and D. Menerey. *Life Cycle Design Manual: Environmental Requirements and the Product System*. EPA/600/R-92/226 Washington, DC, USEPA: 1993.

Kuik, O. and H. Verbruggen, eds. *In Search of Indicators of Sustainable Development*. Dordrecht, Kluwer Academic Publishers: 1991.

Ministry of Housing, Physical Planning and Environment, the Netherlands. *Recyclable versus Disposable: A Comparison of the Environmental Impact of Polystyrene, Paper/Cardboard and Porcelain Crockery*. 1992.

U.S. National Aeronautics and Space Administration (J. Billingham and W. Gilbreath, eds.). *Space Resources and Space Settlements*. Washington, DC, NASA: 1979.

Netherlands Company for Energy and the Environment (NOVEM), The Netherlands Institute for Public health and Hygiene (RIVM), and The Netherlands National Research Programme for Recycling of Waste Materials. *Methodology for Environmental Lifecycle Analysis: International Developments*. Contract Number 8283: 1992.

Repetto, R. "Accounting for environmental assets." *Scientific American* 266(6): 94–100 (1992).

Society of Environmental Toxicology and Chemistry. *A Technical Framework for Life-Cycle Assessment*. Washington, DC, SETAC: 1991.

Stone, R. "Large plots are next test for transgenic crop safety." *Science* 266: 1472-1473 (1994).

Todd, R. "Zero-loss environmental accounting systems," in *The Greening of Industrial Ecosystems*, B.R. Allenby and D.J. Richards, eds. Washington, DC, National Academy Press: 1994.

United Nations. "Integrated Environmental and Economic Accounting." *Handbook of National Accounting, Series F. No. 61*. (interim version) New York, United Nations: 1993.

EXERCISES

1. Select the three sustainability indicators you think are most appropriate for: (a) Sweden, (b) the United States, (c) the People's Republic of China, and (d) the Congo.

    a. Are the indicators you selected different, and if so, why? Justify your selection of each indicator for each country. Consider in your answer state of development, size, population relative to resources, relationship to the global economy (especially trade), existing state of the environment, and any other factors you think relevant.

    b. Given your answer to 1.a., do you think the national state is a relevant political level for the establishment of sustainability indicators? Why or why not? Can you think of any alternatives and, if so, what are their benefits and drawbacks?

2. What should the relationship be between accounting systems and Comprehensive Risk Assessment methodologies? In your answer, be sure to consider the different purposes to which accounting systems at the private firm and national level may be put, and whether these purposes can all be met if you recommend that accounting systems and CRAs should be blended.

3. Many countries are struggling with the policy issues associated with commercial use of transgenic agricultural cultivars. Policies range from strong opposition in some European countries, to fairly restrictive in the United States, at least until recently, to less restrictive in countries such as China.

    a. Assume that you are the Minister of Agriculture for a Northern European nation, and prepare a briefing paper on the subject for your Prime Minister, including your policy recommendations and justifications for them. Include political and cultural as well as economic and technological considerations.

    b. Now assume that you are the Minister of Agriculture for an Asian country with inadequate access to food supplies, and a rapidly growing population. Would your recommendations be different? Why or why not?

CHAPTER 8

Applications to Practice: Research and Development

8.1 WHY RESEARCH AND DEVELOPMENT MATTERS

It appears to be human nature to approach life in two fundamental ways that can be broadly characterized as through skepticism or through faith (Figure 8.1). The skeptical approach answers questions of "why?" through science and questions of "how?"

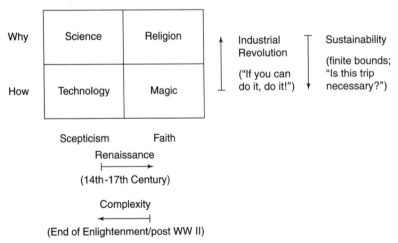

FIGURE 8.1 Scepticism, faith and sustainability. The interplay of scepticism and faith is an important dynamic in understanding the concepts of sustainability and sustainable development, as well as in policy development and deployment.

Based on a model by Doug Merchant, AT&T

through technology, while the equivalents in the realm of faith are religion and magic (all of these terms are used in a non-pejorative sense).

The advent of the Renaissance in Europe began to shift the boundary between these two realms away from faith, which had dominated the West during the Middle Ages, toward skepticism. Subsequently, the Industrial Revolution shifted the boundary to increase the emphasis on "how" and away from "why," particularly as regards technology. It became an industry and market oriented, engineered world.

Recently, however, countervailing trends can be identified. While the power of reductionism in science continues to produce unprecedented advances, it is also becoming more apparent that a number of critical issues only arise at higher levels of systems complexity, and that continuing focus on smaller and smaller systems will inevitably fail to produce an understanding of such phenomenon. To take only a few pertinent examples, the reductionist approach of much toxicology to date is increasingly augmented by efforts to understand ecotoxicity, or the impacts of materials and activities on complex biological communities and not just single species or organisms. The Endangered Species Act in the United States, for example, focuses on individual endangered species, but increasingly ecologists are recognizing that a species focus, while important, must be augmented by a broader understanding of biodiversity within the context of whole ecosystems. And, of course, industrial ecology itself is in part a recognition that the existing regulatory approach focused on individual media, point sources of emissions, and waste sites must be significantly augmented if the goal of policy is not just temporary reduction of anthropogenic risk, but global sustainability.

This raises the second trend that is shifting the boundary away from technology and toward religion: sustainability. As discussed in Chapter 3, sustainability inevitably contains an ethical, subjective dimension; rather than accepting the free market, technological attitude of "If you can do it, do it," it asks "Is this trip necessary?" The latter question has two components: scientific (one must know the implications of taking or not taking a certain step in order to evaluate it) and ethical (by what standards does one determine, based on the data, what is necessary).

The lack of data and scientific understanding regarding virtually all aspects of sustainability has the important effect of making discussions of the concept primarily religious rather than objective. Given the current state of knowledge, when statements regarding "sustainable communities," "sustainable firms," or "sustainable products" are made, they are statements of faith, not science. Thus, for example, when a U.S. Congressman in the mid-1990s remarked that global climate change was "liberal claptrap," what he was really saying was that he regarded global climate change and related issues of sustainability as a myth—a faith that he did not accept. If he had perceived it as a scientific issue, he could have taken exception with the science, but he would not have applied a political, ideological label—"liberal"—to it.

This dynamic explains why research and development is an important aspect of any comprehensive policy structure based on industrial ecology. Figure 8.1 does not suggest that science should replace religion, that either skepticism or faith must dominate. Rather, it suggests that a balance between the two is appropriate, and that, where either is slighted, the other will replace it with suboptimal results. Thus, for example, sustainability taken as a profession of faith may be valuable and motivational, but faith cannot answer the science half of the question, "Is this trip necessary?" To do that, one must have knowledge, and to gain that knowledge, one must have research and development.

8.2 CHANGING RESEARCH AND DEVELOPMENT PATTERNS

In order to discuss the research and development (R&D) structure and needs required for implementation of industrial ecology, it is useful to understand something about the broader context of science and technology (S&T) R&D. This is particularly important because of the fundamental changes occurring throughout the world in this area, expecially in the United States, which has traditionally been a major producer of basic R&D. These are in turn being driven by a number of more basic changes. Table 8.1 lists a number of these trends and their impacts on R&D structures.

Virtually all R&D is carried out by three principle institutions: governments, including national laboratories and nationalized firms, academic institutions, and private firms. All of these institutions are undergoing a period of tumultuous change and redefinition, due in large part to the ongoing so-called "information revolution" and concomitant evolution of the most developed economies toward larger service components and the globalization of world economic activities with the associated continuing evolution of transnational corporations and increase in competitive pressures.

Governments, particularly those of national states, are increasingly devolving political power both upward to international structures and downward to more local

TABLE 8.1 Implications of Global Trends for R&D in Developed Economies

Trend	Manifestation	Impact on R&D Structure
Globalization of national economies	Firms have no protected home markets or product/service lines (minimal rents)	Retrenchment of corporate R&D Refocus of R&D on short term, applied activity
Information revolution, shift to service economy	Private firms and academic institutions changing profoundly and unpredictably	Post-war R&D structure collapsing; future structure unpredictable
Devolution of political power from national state to local governments and transnationals	Profound and unpredictable changes in government structure Loss of centralized coordination of, and responsibility for, national R&D levels and structure	Disrupted R&D Structure
Changing nature of S&T	Increasing importance of "big science" Increasing importance of integrative, multidisciplinary science vs. traditional reductionist, single-discipline approach (i.e., industrial ecology)	Globalization of R&D structures; increased cost of individual R&D projects; increased need for collaboration across cultures; higher transaction costs Primitive infrastructure for supporting integrative science
End of Cold War and accelerating global political and cultural change	Erosion of social consensus, including attitudes towards S&T	Increase in public antipathy towards S&T and R&D Little active support for S&T and R&D

governing units and transnationals. For example, as Edith Brown Weiss points out, between 1918 and 1941, 61 multilateral treaties were recorded; in 1966, the number of international agreements registered with the United Nations topped 33,000, over 900 of which had one or more important provisions concerning the environment. The changing relationships among national-states range from the pressures in Quebec to withdraw from Canada and the violence in Bosnia to the converse trends of increasing regionalization demonstrated by the North American Free Trade Agreement (NAFTA) and the European Union.

Academic institutions struggle to redefine themselves as the Internet and increasing electronic literacy among college age students makes distance learning possible, and changing employment and economic patterns make education a life-long activity, rather than a particular segment of life as a young adult. These trends are exacerbated by cuts in public funding for academic research and related controversy over tenure (life-long employment) for professors as resources become scarce.

Private firms are undergoing a period of rapid change, too, both in response to environmental issues becoming strategic for them and as a result of fundamental upheavals in the global economy. The rate of change of established firms, through merger, acquisition, and bankruptcy, continues to accelerate in response to the pressures of globalization and competition, while some argue that the new vehicle for aggregated economic activity is the "virtual firm," entities that are born, exist, and die in a matter of months. The electronics community in Silicon Valley in California is an example of such a protean system: Is it a series of short-lived firms, or is it in its entirety the new model of the firm?

While interesting in themselves, all of these developments impact the funding and conduct of science and technology research and development (S&T R&D). In particular, they make the perpetual question of how R&D should be funded more difficult. In the United States, for example, R&D has traditionally been funded by two main sources, government and private industry, and conducted in three main types of institutions: government laboratories, private laboratories such as Bell Laboratories, and academic institutions. The trends noted in Table 8.1 have resulted in unplanned reductions in R&D funding from both government and private firm revenues, and, consequently, financial pressure on laboratories, both government and private, and academic institutions. Under these circumstances, funding tends to retract to existing disciplines and programs, and new fields of endeavor, especially integrative fields that lack a disciplinary structure behind them to lobby for funding (such as industrial ecology), are particularly disfavored.

8.3   FUNDING INDUSTRIAL ECOLOGY RESEARCH

From a policy perspective, an initial question must be whether industrial ecology research activity in general is a private or public function. This is a difficult question with no consensus answers as of yet. Traditional R&D categories such as "basic research"—appropriately funded by government—or "applied research"—appropriately funded by private firms—are increasingly less helpful as guides. This is particularly true for a field such as industrial ecology, which is integrative and highly multidisciplinary and cuts across a number of "applied" and "basic" social and physical sciences, as well as virtually all engineering disciplines.

Rather, a more appropriate test is threefold:

1. Are the potential results of the R&D activity useful to society in some way, either through direct economic benefit such as a new technology or direct social benefit such as new knowledge? If so, then the research should be funded. (It is a logical fallacy, by the way, to read this test to imply that research for which obvious potential benefits cannot be identified should *not* be funded: The social benefits of doing such research, at least at some level, are well documented.)

2. Can the benefits of the research reasonably be captured by private interests? This is more than a question of how "applied" the research is: Even research on a specific technology might not be viable in the private sector if the required funding is large and the outcome is too risky to predict a priori. If the results of the research can reasonably be captured by a private firm, then it should be funded from private funds; if, however, private industry cannot foresee or capture the benefit, then there may be a need to fund it publicly.

3. Are there structural reasons why the research, even if desirable, will not be funded by private firms? The most obvious example of a situation like this is an industrial sector that is composed of many small firms, such as the home building or dry cleaning industries in the United States. In these sectors, no single firm is large enough to support an endogenous research capability, and the transaction costs of gathering together so many small firms into any kind of research consortia are prohibitive. Yet, there are significant social costs associated with some of the activities in these firms: Construction activity generates significant material usage and waste and can have large localized environmental impacts such as siltation of surface water; and dry cleaners are a significant source of volatile organic compounds that may be neurotixic or carcinogenic, or contribute to the formation of photochemical smog. Under these circumstances, public support of relevant research and development, even if closely linked to existing technologies and economic interests, might well be warranted.

Using these tests, some practitioners of industrial ecology have argued that much industrial ecology research and development is appropriately supported by government. For one thing, the results of such R&D are public goods, in the sense that they do not benefit any particular private entity (in which case it could be argued that that entity should fund the R&D) but rather result in benefits throughout society. Moreover, such R&D is not only compatible with, but a prerequisite for, proper functioning of a market economy. It is elementary economics that full, or at least reasonably adequate, information is necessary if a free market is to operate efficiently. It is, however, widely recognized, and certainly quite obvious to anyone with experience in DFE or LCA activities, that the vast majority of environmental impacts associated with economic activities and choices are not only still externalities, but in many cases not even identified. Lacking such information, the free functioning of a market economy will inevitably not produce the combination of environmental quality and economic efficiency that is necessary if sustainability is to be approached. Several examples of such projects should make the point clearly.

8.4 ILLUSTRATIVE INDUSTRIAL ECOLOGY RESEARCH AND DEVELOPMENT PROJECTS

Basic research and development in industrial ecology is necessary to provide the objective understanding and support required for the integration of environmental considerations throughout the economy. It is also a necessary prerequisite for the development and implementation of economically and environmentally efficient regulatory structures, currently a critical policy deficiency. What this might entail may perhaps be best illustrated by a few examples; in the next section, a more comprehensive overview of a research agenda for industrial ecology is presented.

8.4.1 Material Stock and Flow Models

These research projects include planning and implementing a series of studies to understand and model stocks, flows, and logistics of material movements throughout the global economy for all major industrial materials, including both renewables and nonrenewables. Some examples of this kind of activity already exist. The World Resources Institute, for example, published a study in 1997 of resource flows in four developed economies: the U.S., Germany, the Netherlands, and Japan. Industrial metabolism studies of materials such as cadmium, lead, and chlorine have also been done, both for regions and globally. Much more of this work remains to be done; it is hindered in many cases by inaccurate data and reporting or simply huge data gaps.

Once they are developed, such models can have environmental impacts and human/ecosystem exposure data mapped onto them, providing the basis for developing environmentally preferable products and processes, and helping industrial sectors and labor markets adjust gracefully to an environmentally preferable world. Such knowledge is also critical to support the development of valid, efficient, risk-based environmental regulations, which is a necessary component of an industrial ecology infrastructure.

8.4.2 Energy System Models

Energy production and use impacts both the social and environmental dimensions of sustainable development. Increased energy production and consumption is necessary for continuing economic development, and at the same time is a principle cause of significant environmental perturbations. The development of integrated "industrial metabolism" models of energy production and use, linked where possible to technology, demographic, and other systems with risk assessment and technology option overlays, is thus a second pertinent example of necessary R&D. Such work facilitates the identification of optimal international, national, and sectoral R&D and investment programs to produce environmentally and economically preferable (and, hopefully, eventually sustainable) energy, manufacturing, transportation, and other technology systems.

8.4.3   Physical Models of Communities

It is increasingly chic for some communities to call themselves "sustainable communities," yet the science to understand what such assertions really mean has yet to be done. Developing integrated models of urban communities, including small, relatively self-contained cities, larger cities with surrounding suburbs, and large megalopolises with decayed centers and most business activity decentralized throughout the suburbs, will be a necessary step in achieving such an understanding. Urban centers in developing and developed nations should also be modeled and compared. Such models would include transportation, physical infrastructure, food, energy, material stocks and flows, and other systems. Both direct and embedded impacts (e.g., impacts caused by the manufacture of products or growth of food which the city imports) should be included. This information would facilitate identification of major sources of environmental impacts, patterns of activities that give rise to them, and potential environmentally preferable technological or mitigation options. It would also provide a necessary basis for comparing the environmental impacts of different kinds of communities, as well as creating higher level, integrated regional models.

8.4.4   Sector Models

Achieving greater economic and environmental efficiency will have substantial, and differential, impacts on existing economic sectors. Integrated models of specific sectors (on both national and international levels) of particular economic, environmental, or cultural importance—including, for example, the agriculture, forestry, extractive, electronic, and automotive sectors—could be used to understand how they might be affected by an increasingly environmentally sensitive world. Such research could be particularly important in mitigating potential economic and employment shocks of discontinuous environmental and/or related economic and regulatory changes, and in supporting continued improvement in quality of life while reducing attendant environmental impacts.

8.4.5   Technological Evolution

One of the more robust hypotheses of industrial ecology to date is that rapid evolution of environmentally appropriate technological systems is a prerequisite for improvement of quality-of-life in an environmentally sensitive world. As Richard Nelson of Columbia University points out in his book *The Sources of Economic Growth*, the fundamentals of technological evolution and diffusion throughout the economy are, however, poorly understood. It is not clear, for example, what optimum or maximum rates of technological evolution might be; what associated economic and labor costs and benefits might be (and how they could be optimized); and how such variables differ by class of technology and state of development of the national economy.

Nor are the meaningful barriers, both cultural and scientific, to rapid technological evolution easy to identify. For example, it is apparent that moving to a hydrogen-based energy economy will be significantly more difficult and a far more lengthy process than was substituting for CFC-based cleaning systems in electronics manufacture. Some reasons are intuitively obvious: potential opposition from the petroleum industry if it expects to lose a large market, for example. Others are more subtle: The degree to which a technological change is linked to other technology systems and infrastructure, and thus requires changes in those systems rather than just itself, is obviously important, as is the cultural context within which the existing technology is used.

8.4.6   Industrial Ecology Policy Studies

Investigating the interdependency of legal, economic, cultural, marketing, scientific, and technological activities and policies as they affect environmental protection and the evolution of environmentally appropriate technological systems is necessary if policies supporting sustainability are to be developed. Studies of different regulatory tools and approaches in terms of how private firm and consumer behavior subsequently shift, for example, could be quite useful in developing efficient private and public environmental management structures.

8.4.7   Data Management and Integration

Vast amounts of data exist, but they are essentially inaccessible in files and databases across all levels of government, including both developed and developing nations. They are also of widely varying quality, from formal toxicology studies to second hand anecdotal allegations of impacts from various substances. These data resources should be identified, and the data compiled and controlled for quality. They should then be translated into tools that make them easily accessible to entities that can use the data to improve their environmental performance. This offers an easy and important mechanism for beginning industrial ecology research without requiring substantial new resources, which in many countries will not realistically be available in the foreseeable future.

8.5   AN OVERVIEW OF INDUSTRIAL ECOLOGY RESEARCH REQUIREMENTS

Figure 8.2 illustrates an industrial ecology research structure. Across the top are analytical units, or types of systems that need to be understood with some rigor if discussions of approaching sustainability are ever to be more than verbiage. Down the side are the types of analyses that should be done for each system. The intersection of these categories creates a cell. Each cell is a research program in itself, and no cell has, to date, even a partially adequate body of work behind it. More than anything else in this text, this figure demonstrates the profound depths of current ignorance about industrial ecology. It should also be noted that this is a preliminary effort based on the current state of understanding of industrial ecology, and so, in itself, is undoubtedly incomplete and will be elaborated on as the field advances.

8.5.1   Analytical Units

These analytical units or types of systems are those that must, at a minimum, be understood to begin to identify in a systematic manner the links between human economic activity and underlying natural systems. They are in many cases related, but each unit

Industrial Ecology Research Structure

Analytical unit (system) / Type of analysis	Materials				Energy			Products		Scaled systems			
	Grand cycles	Renewables	Non-renewables	Water	Fossil	Nuclear	Renewable	Simple ↔	Complex	Land use / Services / Sectors		Economic / Geographic / Temporal / Social	
Stocks and flows													
Inherent hazards													
Exposures													
Risk/cost/benefit				(Cells connected by integrated system of metrics)									
Technology options													
Economic factors													
Policy options													
Social and cultural implications													

FIGURE 8.2   An illustrative Industrial ecology research agenda. Each cell in itself represents a significant research challenge.

represents an important separate set of data and an analytical entity, which must also be understood on its own.

The need to understand the life-cycle impacts of individual materials and energy systems, as well as those systems as a whole, is apparent. Particular attention to the so-called "grand cycles" of carbon, nitrogen, sulfur, and phosphorous, particularly as they have all been perturbed by human economic activity to a meaningful extent, is obviously important. Of the other materials, water and nuclear materials are also worth emphasizing, albeit they raise far different issues and have very different cycles.

It is also necessary to understand products taken as a whole. Here, however, there is an important difference between simple products and complex products. *Simple products* are those whose function depends primarily on their material composition, such as food, motor oil, shampoo, or pesticides. In many cases, simple products involve dispersive uses (that is, the use of the product necessarily disperses it into the environment). *Complex products*, on the other hand, derive their value not from their material composition, but from the intellectual content embedded in their design; examples include personal computers, airplanes, automobiles, and white goods. The environmental role of a complex product in society is usually far harder to capture than that of a simple product, in part because complex products are frequently expressed in the economy as services. This difference between product categories, while generic and seemingly obvious, has not been adequately recognized by some in the environmental community who, for example, attempt to apply tools such as LCA, which are perhaps adequate for simple products, to complex products, where they simply don't apply. For example, if one knows the potential environmental impacts of the principle ingredient of a shampoo, one can predict with some validity the environmental impacts of that product. If one knows the environmental impact of aluminum, on the other hand, one still knows virtually nothing about the environmental impacts of a Boeing 747.

The environmental impacts of the service sector, although frequently overlooked, can be quite substantial. The use of air freight to deliver mail order packages, for example, embeds a fair amount of energy in that product by the time the customer gets it. Impacts of postal systems, hospitals, retail operations, and malls are all different and potentially significant. Both direct impacts and indirect impacts need to be researched. Retail operations, for example, have a number of direct impacts—demand for transportation, facility construction and maintenance, maintaining inventory—but their indirect impacts—stimulating automobile traffic, encouraging high levels of material consumption—may be more difficult to manage. More importantly, evolution toward the "functionality economy" popular among European industrial ecology thinkers, where customers purchase a product's function rather than the product itself (e.g., leasing computers, automobiles, white goods, and other complex products), implies a growing and environmentally more efficient role for service activities in the economy.

Expanding that notion, it is apparent that evolution of the global economy toward sustainability will differentially impact different sectors. For example, the energy and mining sectors may have some difficult adjustments to make, while the electronics sector, particularly software providers, and the communications sector may well profit by such an evolution. Moreover, one can anticipate that the boundaries between sectors that are now taken for granted will become much more fluid. Mining, chemical, and waste management companies may all find themselves in competition to provide material management services to manufacturers that will be leasing on both the supply and demand sides of their operations: They will lease materials to incorporate into products, concomitantly developing close relationships with their material management suppliers, while at the same time leasing their product to consumers as part of the functionality economy. Agricultural technology may increasingly serve as a production base for the previously completely separate pharmaceutical sector. Obviously, such shifts will have significant implications for employment levels and requisite skill sets in the future; research in this area might have as one goal the minimization of transitional adjustment costs to workers and national economies.

The "scaled systems" category in Figure 8.1 represents four different types of complex systems that operate at many different scales and exhibit a number of different behaviors at different scales and in their interrelationships. Economic systems range from the family as an economic unit, to local communities and intranational regions, to national states, to regional entities such as NAFTA and the European Union, to the global economy. Geographic systems also range from communities to the global level, and are somewhat correlated with social systems. The latter, however, are becoming less geographically defined as communication systems such as the Internet become more ubiquitous: For increasing numbers of people, one's social set is increasingly defined by interest, not by geography.

Temporal systems range from the immediate acute effects of an emergency spill, to chronic effects of exposure to toxic materials, which can express themselves over the life time of an organism, to century long perturbations of atmospheric, oceanic, and biological systems. Understanding temporal systems in their entirety is important because, while many people, and thus political systems, tend to focus on the immediate (e.g., cleanup of hazardous waste sites), it is the longer term perturbations, such as global climate change, that may have the greatest economic, social, and biological impacts, and are the hardest to mitigate.

Each of these scaled systems must be understood at each relevant level. It is equally as important, however, that the relationships between levels be understood as well. The appropriate mental model is a network structure, where both the nodes and their inter-relationships must be understood if the network as a whole is to be comprehended.

8.5.2 Types of Analysis

A number of different kinds of analysis should be performed for each type of system if it is to be properly characterized and understood; those identified in Figure 8.2 should be regarded as illustrative rather than definitive. Obviously, one of the most essential types of analysis is to identify and model the physical basis for the system, its stocks and flows. Once this is done, the inherent hazards present at different points in the system can be identified. Identifying the exposures facilitates the generation of a comprehensive risk assessment across the system, which should include economic factors so that the costs and benefits of different options, and potential mitigating activities, are apparent. Obviously, a systems-based approach is required here to ensure that an activity taken to reduce one impact does not generate unacceptable impacts elsewhere in the system.

Technology options should also be identified throughout the system, and should include not just existing technology, but that which can be relatively easily developed and deployed. Where identified impacts and risks are substantial enough, new programs of targeted technology research might be warranted. In general, the developing understanding of the system should encourage a technology research effort aimed at efficient improvement of the environmental and economic dimensions of the system.

It is obviously desirable that policy development be based so far as possible on the scientific and technological characteristics of the system, and with an understanding of the implications for system performance in both the short and long term. Unlike technical systems studies, however, policy development must also be sensitive to social and cultural dimensions of the issues involved, and thus represents, in a sense, the synthesis of the scientific and technological with the social and cultural. Policy development, in other words, must be systems-based, and is the stage at which sustainability, rather than just the objective study of industrial ecology, becomes the manifest purpose.

8.5.3 Metrics

Analogous to the need for metrics to link different aspects of the industrial ecology conceptual framework (Chapter 7), sets of metrics are required so that, even as each complex system is studied in itself, the relationships among these systems and the emergent behavior that is displayed as they are integrated does not become lost. Metrics, in other words, define and support the inter-relationships among the nodes in the industrial ecology research network structure. Thus, for example, it would obviously be appropriate to have metrics that link materials stock and flow patterns to sectors: If a particular material were a problem in a particular application, substitutes from other sectors might be used to replace it, which might require different technological options.

Using energy as another example, the use of fossil fuel might be reduced by developing microbial systems that digest biomass into methane, which might require biotechnology development. If such digestion could occur locally, one could envision the beginning of an evolution away from global petroleum distribution infrastructures toward local production of biomass-based petroleum substitutes. Integrating energy and materials assessments, plastics could be made from such renewable hydrocarbon resources, used, and then burned for energy recovery without contributing to global climate change forcing.

8.5.4   Evaluating a Small Community

To illustrate the concept of such an industrial ecology assessment, consider a small community in the United States. The first step in understanding it from an industrial ecology perspective would be to model the stocks and flows of energy and materials that underlay the structure of the community. In many cases, this distributes the potential environmental impacts of the community beyond its borders: Petroleum will have been pumped, refined, and shipped from elsewhere; manufactured products will have created emissions elsewhere; agricultural materials and food will have generated their ecological impacts elsewhere. Land use patterns, again both explicit and implicit (that is, embedded in products and energy imported into the community), are an integral part of mapping the community's physical basis.

Once the physical basis for the community is established, the environmental impacts created across the broad scale of geographical and temporal systems can be identified by taking an inventory of the inherent hazards embedded and explicit in the material, product, and energy stocks and flows, and the exposures that, taken with inherent hazard, generate risk. These findings can then support a broader economic and technological assessment of options, generating, if desired, policies that can focus mitigation efforts on the most pressing and meaningful concerns. These policy options, of course, will integrate social and cultural implications; otherwise, they will not be viable politically. For example, if automobile emissions turn out to be a major contributor to the environmental inefficiency of the community, programs to turn in older cars or catch particularly "dirty" heavy emitters might be implemented. Plans to convert the community to hydrogen vehicles, on the other hand, obviously would be difficult to implement in the short term in most areas.

There are several instructive points about even this brief conceptual sketch. Most importantly, the methodologies or the data to do it are at an exceedingly primitive stage. It is not just that incremental work is needed to generate the product, it is that the right questions are just beginning to be asked. The second point is the obvious links among different kinds of systems: The impact of the automobile on the community cannot be assessed without understanding something about that technology system, and the same is true of energy, agriculture, raw materials, finished products, and all the other elements relevant to the community system.

Even this simple example also demonstrates the fallacy of those private firms, communities, and products that bill themselves as "sustainable." Leaving aside the problem that sustainability remains an ambiguous concept, one is immediately struck by the fact that such entities cannot be isolated from the other systems with which they interact, and so cannot be meaningfully called sustainable without integrating their effects, which in turn depend on other systems in other places and other times. Firms, products, or even nations may be more or less environmentally efficient, but sustainability is a global property, and cannot be subdivided. It either exists as a property of global human economic activity over time, or it does not.

8.6   EARTH SYSTEMS SCIENCE AND ENGINEERING

As discussed earlier in the introduction to this text, natural global systems of all kinds are increasingly being altered—unconsciously engineered—by human activities. This long-standing trend raises the prospect of a challenging scientific and engineering research agenda focused on human interaction, and eventually governance, of many fundamental natural systems. What is being done haphazardly, dangerously, and in ignorance now may in future years become an integral part of a mature human relationship with the planet itself. This is the concept and research field of earth systems science and engineering.

An example might help clarify this type of research. If there is one sector that is almost universally regarded as being an enemy of the environment, it is fossil fuels. And yet, technology exists or is foreseeable that offers the possibility of migrating the fossil fuel sector from environmental villain to critical carbon cycle governor. If existing carbon sequestration technology, which can capture the carbon dioxide from a fossil fuel power plant for deposition in long term sinks (geologic or perhaps oceanic), is combined with substitution of hydrogen for fossil fuel in land transport systems, which could be done with existing hybrid vehicle technology (combining a hydrogen internal combustion or fuel cell engine with battery or flywheel power storage systems), one could envision a primarily fossil fuel-based energy system that contributes little to global climate change or other environmental impacts (Figure 8.3).

The basic model would be that large centralized fossil fuel power plants, located near fossil fuel sources and sinks (which might be the same geological formation, for example), would produce electricity for distribution over the grid and hydrogen for mobile energy uses, primarily transportation. Depending on economic and technological considerations, the hydrogen might be shipped in bulk in chemical form (hydrides, for example), or it might be shipped through pipelines, either as itself or as methane for final processing and carbon sequestration closer to the final use point. Energy production facilities would be designed to capture and sequester not only carbon, but whatever other problematic species were produced (capture of mercury, other heavy metals, and radionuclides from coal-fired plants would be one obvious priority).

In the longer term, however, such plants could be designed to accept mixed biomass-fossil fuel, and carbonaceous municipal waste inputs, thus turning the carbon-based energy system into an anthropogenic carbon cycle governor, rather than a source of potential global climate change forcing. Two functions, the ratios of biomass to fossil fuel and carbon sequestered to that released, could be modulated to produce

The Carbon-Based Energy System as Carbon Cycle Govenor

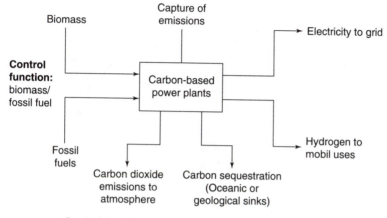

Control function: CO_2 sequestered/CO_2 emitted

FIGURE 8.3   Schematic of a fossil fuel power plant component of a carbon cycle governor system.

over time the appropriate level of carbon dioxide in the atmosphere. Obviously, carbon sequestration would only be part of a more complex engineered carbon cycle, which would undoubtedly include many technologies and components, but it is a useful illustration of the concept. The fundamental policy approach would shift from not recognizing or accepting the impacts of human activity on the carbon cycle to accepting it and taking responsibility for managing it properly.

Obviously, earth systems engineering approaches such as this could be broadened; one can see agricultural practices, for example, being used to modulate the nitrogen cycle, although this would be extremely difficult simply because agriculture will be a diffuse, atomistic activity for the foreseeable future. But it is important to recognize and emphasize that the current state of knowledge is nowhere near adequate to support such activities yet. Knowledge of these complex systems and their dynamics is far less than that which would be required to move ahead responsibly. The carbon scenario, for example, raises questions of risk and technological evolution that have yet to be properly phrased, much less resolved (e.g., how safe are the carbon sequestration methods currently identified?). Equally as fundamental, neither the institutional nor the ethical basis for undertaking such activities currently exists. It raises difficult moral, institutional, scientific, and technological questions to suppose that a human global society will be able, or should presume, to modulate fundamental natural cycles as a deliberate engineering exercise.

On the other hand, it is apparent that human activity is already profoundly affecting many natural systems, such as the carbon cycle. Recognizing this point, some have argued that it becomes difficult, if not unethical, to continue to plead ignorance as an excuse for not taking responsibility, and learning how to exercise that responsibility appropriately.

The last point bears elaboration. Responsibility is both ethical and scientific in such cases: what is desired (and how is that desire measured when values of different groups may conflict), and what data and understanding are required to carry out that desire? Policy development must constantly avoid superficial, and irresponsible, suggestions. For example, at one point when concern about stratospheric ozone depletion as a result of releases of chlorofluorocarbons from electronics manufacturing was rising, a suggestion was made, apparently in all seriousness, that tropospheric ozone (smog) simply be pumped into the stratosphere and all would be well. Both the science and common sense strongly argued against even considering such a course. On the other hand, recent experiments have demonstrated that iron fertilization of oceanic systems can stimulate the growth of phytoplankton, which as a result take up carbon, a process which, if conducted at large enough scale, might help mitigate global climate change, and thus be another part of an engineered global carbon cycle. The systems involved are clearly complex and not well understood, however, so proceeding too rapidly at too large a scale at this point would be considered irresponsible by most people. Nonetheless, the possible risks of global climate change are also non-trivial, so research exploring this option is not inappropriate.

REFERENCES

Adriaanse, A., S. Hringezu, A. Hammond, Y. Moriguchi, E. Rodenburg, D. Rogich, and H. Schutz. *Resource Flows: The Material Basis of Industrial Economies.* Washington, DC, the World Resources Institute; Wuppertal, Germany, the Wuppertal Institute; The Hague, Netherlands, the Netherlands Ministry of Housing, Spatial Planning and Environment; and Tsukuba, Japan, the National Institute for Environmental Studies: 1997.

Allenby, B.R. "An industrial ecology research agenda." *Pollution Prevention Review.* Winter 1998.

Eisenberger, P.M., ed. *Basic Research Needs for Environmentally Responsive Technologies of the Future.* Princeton, Princeton University Materials Institute: 1996.

Nelson, R.R. *The Sources of Economic Growth.* Cambridge, MA, Harvard University Press: 1996.

Smith, B.L.R. *American Science Policy Since World War II.* Washington, DC, Brookings Institution: 1990.

Weiss, E.B. "The changing structure of international law." Inaugural Lecture, Francis Cabell Brown Professorship of International Law, Georgetown University Law Center, May 23, 1996.

Wernick, I.K., J.H. Ausubel, and the Vishnus. *Industrial Ecology: Some Directions for Research.* (Prepared for the Office of Energy and Environmental Systems, Lawrence Livermore National Laboratory.)

EXERCISES

1. Select one simple product and one complex product. Identify and describe (a) the main environmental impacts they create over their life cycle, (b) potential substitutes for the functions they fulfill, and (c) the economic, social, and technological barriers to successful substitution of alternatives. Which product is most easily replaced? Which product is more embedded in the economy and in society? Which barriers to substitution are most difficult to overcome?

2. You are the Technology Minister for a small, urbanized Northern European country. Your country has just adopted sustainability as a national vision, and you now wish to understand what that may mean in practical policy terms. What research would you initi-

ate to begin answering that question? What research would you anticipate private firms conducting in response to the national vision? How would you encourage a joint industry/government research program?

3. Five years ago, the university at which you teach decided to become a "virtual university," and now 90 percent of its students are distance learners, and all the campus buildings but the Computer Center have been sold to a raging capitalist for conversion to condominium units. You have been tasked with developing a research program in industrial ecology. Prepare a two-page outline of your plan, including funding sources, personnel requirements, data sources, and organizational structure of the research activity.

PART II

INDUSTRIAL ECOLOGY POLICY DEVELOPMENT

CHAPTER 9

Complex Systems

9.1 WHY ASSUMPTIONS ABOUT SYSTEMS MATTER

The subject of systems has already been introduced in the discussion of industrial ecology, a field that is predicated on a recognition that the regional and global environmental perturbations of greatest concern cannot be substantially mitigated without adopting a systems-based approach. A critical element of industrial ecology, in fact, is to understand how economic systems can be evolved so that materials do not simply pass through the economy in a linear fashion and become waste (a Type I system), but are cycled within the economy to the greatest practicable extent (approaching a Type III system).

There is another important dimension to this systems focus, however. If efficient and effective policies aimed at implementing industrial ecology and fostering the evolution of the global economy toward sustainability are to be developed, it is also important to understand the differences between *complex* and *simple* systems, and the implications for policy formulation that flow from these differences. These two classes of systems behave in quite different ways, and are perceived by people in quite different ways as well.

Although there are interesting cultural variations, human psychology itself appears to predispose people to think in terms of simple systems, conceptualizing in terms of relatively few variables whose interrelationships are easily and relatively completely understood, and that are displayed over a short time period. Many aspects of complex systems are, on the contrary, difficult and counterintuitive, and frequently are illustrated only by the behavior of properly constructed quantitative models.

It is critical to recognize, however, that adopting the industrial ecology approach, and sustainability as a goal, require that we work with complex systems: The shift in policy endpoint from localized reduction in (usually human) risk implicit in much environmental regulation to date, to a goal of global economic and environmental sustainability carries with it a fundamental shift in focus from simple to complex systems. Reducing human risk from a specific site or a specific set of emissions, while it may be quite a complicated process in practice, is conceptually simple. Ideally, it can be accomplished by a relatively simple regulation: a clean-up order, for example, or a ban. The simple endpoint tightly bounds the scope of risk assessment, and the consideration of possible trade-offs.

Sustainability, however, may well be an emergent characteristic of a properly functioning global economy, and that is quite a different matter. It implies a relationship among all economic activities, and between them and the relevant environmental systems that, taken as a whole, achieves a state (sustainability) that we cannot yet define. Moreover, because sustainability is a property of the economic/environmental system as a whole, policymakers cannot afford the luxury of ignoring the many impacts of their actions, for it is the summed impacts, intended and unintended, that determine whether a regulatory intervention has advanced, or even retarded, the approach toward sustainability. An understanding of systems, particularly of complex systems, is thus critical to rational industrial ecology and sustainability policy formulation. This discussion, therefore, covers the most pertinent aspects of complex systems; it does not discuss the mathematics of their structure or behavior, however, which can be quite difficult. Moreover, at this point it is not necessary to link this systems discussion with the "science of complexity" trend; while some of the principles and models being developed in that area may indeed be applicable to these issues, it suffices for now to provide the reader with well understood characteristics of different classes of systems, and tease out their major implications for research and policy. Readers desiring advanced treatment of these concepts can refer to some of the references cited at the end of this chapter.

9.2   SIMPLE AND COMPLEX SYSTEMS

While considerations of systems and their complexity can be quite complicated, the basic concepts are relatively simple and easy to grasp, although their implications, especially for policy, are less straight foreword. *Systems* may be defined as groups of interacting, interdependent parts (agents, or subsystems, or lower hierarchies within the system under consideration) linked by exchanges of energy, matter, and/or information. Generally, systems are delineated for analytical purposes by a boundary that separates the internal workings of the system from its external environment. Drawing such boundaries is a matter of convenience and the purpose for which a system is being defined, and is inevitably somewhat arbitrary. In general, a good boundary is one that internalizes as much of the relevant interactions as possible, while at the same time minimizing the relevant interactions of the system with its environment, thus facilitating analysis.

Bounding the system to be considered is important. It is also critical, of course, to understand the nature of the system as thus defined. The most fundamental difference between types of systems for purposes of industrial ecology is the difference between simple and complex systems. This is a critical distinction because traditional reductionist approaches common to economics, environmental science, and other disciplines do not provide a good basis for understanding complex systems, and can lead to substantial policy problems (such as environmental regulations that are, on balance, not just economically inefficient, but actually bad for the environment). Table 9.1 lays out the most pertinent differences between complex and simple systems, and their implications for policy. The most fundamental of these is elementary, but profound in its implications: Simple systems are intuitively "understandable," and complex systems are not. This arises from several key differences between these two classes of systems.

TABLE 9.1   Policy Implications of Simple (S) versus Complex (C) Systems

Function	Type	Function as Displayed by System	Policy Implication
Information	S	Centralized; system is "knowable"	Centralized command-and-control feasible
	C	Information diffused throughout system; some embedded in system structure; system too complex to be "known"	System management by adjusting forcing functions and observing changes in behavior; command-and-control contraindicated
Causality	S	Linear; cause and effect easy to determine	Centralized command-and-control to endpoint (effect) feasible
	C	Causes and effects cannot be linked in most cases	Cannot be sure of actual impact on system of any policy initiative
Response to Forcing	S	Predictable and relatively linear	Rigid regulatory structures O.K. (because outcome predictable) and politically preferable (because can't be gamed)
	C	Highly non-linear and may be discontinuous; not predictable a *priori*	Policy once in place must be very flexible, so it can be changed as system response dictates: arbitrariness and gaming are avoided by the adherence to an appropriate set of metrics
Control Mechanisms	S	Rational centralized control is possible and effective	Single, fully responsible entity with authoritarian power can control system (e.g., U.S. Environmental Protection Agency)
	C	Control nodes are diffused throughout system; centralized control impossible because of a lack of information and unpredictability of system response	Responsible management entity adjusts forcing functions (e.g., taxes or fees) and monitors results; direct control of system towards endpoint not possible
Endpoint	S	Defined in static terms and achievable, not path dependent	Command-and-control management to endpoint O.K.
	C	Defined in process or performance terms and not fixed because of constantly evolving nature of system	No defined endpoint; manage system to achieve desired emergent behavior
Metrics	S	May not be necessary because performance is measured by attaining endpoint, not system state	Endpoints are metrics
	C	Defined in process or performance terms and not fixed because of constantly evolving nature of system	Manage process to improve performance against metrics; metrics critical to public/political acceptance of management process; choice of appropriate set of metrics is essential to policy evaluation and subject to improvement
Existence	S	Produce defined endpoint	Can minimize specific insult (e.g., lead dispersion from use in gasoline); cannot lead to or support achievement of sustainability
	C	Maintain system stability and integrity over time	Policy focus as a whole must be on the system, not on arbitrary components; metrics focus on stability and systems' dynamics, not throughput

9.2.1 Linearity

Simple systems tend to display linearity between inputs and outputs. If an input changes by a certain amount, the output changes by an equivalent and predictable amount; moreover, the change in output is relatively comparable over the entire range of inputs and outputs. Complex systems, on the other hand, are generally characterized by strong, at least partially non-linear interactions between the system components. If an input changes by a small amount, the output may change dramatically depending on the internal system dynamics. For example, a salt marsh may look relatively unaffected for a long time as the pollution input increases, but at some point just a small increase in pollution results in a significant degradation in the salt marsh ecology. Global climate change forcing may have no obvious effects on oceanic systems for a while, but at some point there is a reasonably high probability that ocean currents will shift suddenly, leading to a discontinuous change in European climate. These are non-linear responses.

9.2.2 Causality

It is usually easy to determine cause and effect relationships in simple systems. If a factory takes in a certain amount of raw materials, for example, at some point it will produce a corresponding amount of product and waste. Complex systems, however, tend to be characterized by complex feedback loops that make it difficult to distinguish cause from effect. Thus, given a biological community such as a salt marsh, inputs and outputs may be very difficult to predict, especially if the system is stressed and thus likely to shift non-linearly to a new state. Species composition, interactions within and among species, competition patterns and feedback on species numbers, shifts in material cycles, resiliency of the system's state to further stressing: All make it difficult to predict exactly how the system will react, or, indeed, exactly how to define the system in the first place. The causal links within the community, and between the community and its surrounding environment, are far less evident than in the case of the factory, which has been engineered to be a simple, predictable system. In such cases, cause and effect becomes an inappropriate way to think about the system's function.

This example, however, makes another important point: Depending on the purpose, one may look at the same set of interrelationships and conceptualize it as a simple, or as a complex system. These definitional gradations among types of systems are for our analytical benefit, and thus to some extent depend on the purposes for which the system is being considered. In either case, of course, the underlying structures, functions, and relationships are the same; it is the mapping to a model which changes.

9.2.3 Lags, Discontinuities, Limits, and Thresholds

Simple systems tend to be intuitive in the sense that their operations are all apparent. In practice, this means that there are no significant spatial and temporal lags, discontinuities, or limits and thresholds. Most people think of global climate change in this manner, and thus tend to implicitly assume that, because the forcing functions are changing only gradually, and temperature increases to date have been only gradual, any future changes will also be gradual. Moreover, the lag times involved—decades to centuries before both increases and precautionary decreases in forcing functions have full effects on the climate system—are beyond the intuitive understanding of most individuals and most social and political institutions. Because global climate patterns and the oceanic and biological systems with which they are integrated are complex systems, this intuitive comfort is somewhat dangerous, as it is almost surely wrong. Complex systems generate internal behavior seemingly "out of nowhere" as a result of their internal dynamics: Spatial and temporal lag times cover a broad range, and cannot be adequately predicted simply by focusing on those near term events that can be easily predicted. Complex systems are also known for discontinuities, where a small change in input creates a discontinuous change in output, and frequently display a number of limit and threshold effects. In most such cases, these types of responses from complex systems cannot be predicted. Accordingly, people and policies tend inevitably to assume they don't exist, a dangerous and usually false assumption.

9.2.4 Process as Endpoint

Simple systems tend to have a defined, static equilibrium point toward which they move regardless of starting point and independent of time. A marble in a bowl, for example, will move toward the bottom of the bowl regardless of where on the bowl's surface you start it or when you let go of it. Complex systems, on the other hand, usually have no such easily defined equilibrium point. Rather, they tend to operate far from any equilibrium point(s) in a state of constant adaptation to changing conditions. Modern views of ecosystems, for example, tend to eschew the traditional view of ecological succession (communities replacing or succeeding prior ones in the same location) as a "progression" to a specific stable endpoint, which is then maintained forever through time. Rather, the community constantly shifts in response to its internal dynamics and external environment, which is in turn constantly shifting. The neat progression towards finality is replaced by continual process: Being is integrated into forever Becoming.

9.2.5 Emergent Behavior

It is characteristic of systems that they contain a hierarchy of levels, each including the levels below them. For example, remember the automobile system illustrated in Figure 4.9: The engine system was one level, consisting itself of lower levels (components, materials, subassemblies, and so on); a number of such systems is fitted together to make an automobile; the car in itself is a component of a larger transport system, which includes built infrastructure, energy delivery systems, and a number of other subsystems, each in turn being composed of a number of its own subsystem levels.

Simple systems contain no surprises as one moves up the level of system hierarchies. Properties of system levels can be predicted and understood just by adding up the properties of the constituent subsystems. Thus, for example, the composition of a fruit salad can be understood simply by adding up the properties of the different fruit that were cut up to make it.

Complex systems, however, are generally characterized by emergent characteristics as one moves from subsystems to higher levels in the system hierarchy; that is, behavior that occurs only once that level of the system hierarchy is reached, and cannot be obtained simply by summing up the contribution of the parts of the system to the whole. Thus, for example, a cell exhibits behavior that could not be predicted simply by knowing its chemical constituents, and human consciousness would be impossible to predict simply through the study of cells. Similarly, one could not predict the behavior of an ant hill simply by observing an ant. In each case, the behavior that emerges at the higher level is qualitatively different than what could be predicted by considering the characteristics of the lower levels alone.

The concept of emergent behavior is particularly relevant to our subject. If, for example, sustainability is a high-level emergent characteristic of a properly bounded and functioning complex system (i.e., a sustainable global economy), then it is unlikely to be properly understood by traditional, reductionist approaches. It follows that policies designed to achieve sustainability that are based on such approaches will likely be fundamentally incomplete or flawed.

9.2.6   System Evolution

Simple systems do not evolve. If disturbed, like the marble in the bowl, they return to equilibrium. In contrast, all complex systems evolve in response to changing boundary conditions and internal dynamics. Evolution occurs in a process involving three factors: (1) an information storage and transmission mechanism, which is subject to and can capture (2) mutations, which are simply new alternatives for the agents within the system, and are themselves sieved by a (3) selection process reflecting internal states and external boundary conditions. Thus, for example, an organism has its DNA as an information storage medium, which can, under a number of circumstances, be altered to create mutations, which then provide the information for generating a new organism. This representation of the changed genetic code is then tested in its environment, and either survives to reproduce or dies. A commercial technology is a representation of existing information, while technological innovations are equivalent to mutations. These are indeed sieved by the market, and the complex of legal, political, cultural and ideological systems linked with any economy in practice. In both these instances, however, the other aspects of complex systems cannot be forgotten. In particular, the history and boundary conditions of the external environment, the complex dynamics of the mutated entity with its external environment, and simple chance will play roles in the eventual evolution of the system to either include or exclude the new agent. Nothing is inevitable or optimized as a complex system evolves, especially if, as in the case of individuals, firms, or society, there is an element of (bounded) free will embedded in the system's behavior.

The dynamics of evolution, in turn, result in quite different system behavior than that traditionally assumed in economic optimization analysis. For one thing, the evolution of a complex system is path dependent: History matters. For another thing, because there are no "optimal" results, evolution can achieve multiple equilibria. There is also no assurance of optimal efficiency in an evolving complex system, because its state is dependent on its history and is sensitive to internal and external perturbations. Finally, survival of the first, rather than survival of the fittest, is a likely outcome. Static equilibrium analysis does not adequately explain the dynamics or behavior of such systems.

9.3   INTEGRATING SYSTEM LEVELS IN INDUSTRIAL ECOLOGY

Integrating systems levels in industrial ecology given the complex nature of the system presents a challenge to the policymaker desiring to implement industrial ecology principles, or, indeed, any entity seeking to advance toward sustainability. On the one hand, unknowingly mixing different levels of the hierarchy will result in a confused, unintelligible program. Thus, for example, an improvement in process chemistry that increases economic and environmental efficiency is desirable, but should not—indeed, cannot—be justified by references to sustainability. The product being manufactured by this more efficient process, for example, may be used primarily to cut down rain forests. More fundamentally, if sustainability is indeed an emergent characteristic of a properly organized global economic/environmental system, it is impossible to say that anything is sustainable at a lower hierarchical level than that of the whole system itself. A claim to be a "sustainable firm" or to make a "sustainable product" is not a meaningful statement.

Nonetheless, over time it is obviously important that broad industrial ecology principles inform individual design decisions and that they, in their turn, inform industrial ecology research activities. In other words, system levels should not be confused, but they should communicate. This is accomplished through the development of systems of metrics, which, if properly designed, capture the information in one level of the system required to ensure appropriate performance at another level of the system (see Chapter 7 for a discussion of metrics). The mental model for complex systems is an ever-changing network of networks, not a static, separate object: the Internet, not a factory stamping out widgets.

Understanding relevant levels of complex systems and how they are integrated is not a new problem. Ecologists and climate modellers, for example, have had to struggle with analogous issues as they attempt to link ecological studies, performed at the scale of hectares, with general circulation models (GCMs), which can have grid scales of 500-by-500 kilometers. Root and Schneider have suggested a methodology called "strategic cyclical scaling," or SCS, which deliberately cycles between scale levels, linking research design and activities, to be used in such instances. A similar system can be established for the industrial ecology systems hierarchy as well, as suggested by Figure 2.1. From a policy perspective, this helps clarify how information and research can be focused, both within each level and between the levels, and unnecessary confusion because of "level-jumping" can be avoided.

9.4   POLICY IMPLICATIONS OF SYSTEM STRUCTURE

In discussing the prevalence of organizational failure in modern industrial entities, Senge and Sterman of the MIT Sloan School of Management make the pertinent point that humans are poor at managing complexity:

> For systems theorists, the source of poor performance, organizational failure, and inability to adapt is often to be found in the limited cognitive skills and capabilities of individuals compared to the complexity of the systems they are called upon to manage. A vast body of experimental work demonstrates that individuals make significant, systematic errors in diverse problems of judgment and choice .... Dynamic decision making is particularly difficult, especially when decisions have indirect, delayed, nonlinear, and multiple effects. Yet

these are precisely the situations in which managers must act. The turbulence of the late 20th century is in large measure due to increasing complexity of feedbacks among institutions and our inability to understand the dynamics they generate . . . . Throughout these studies [of management failure to understand and manage complex systems] runs a common theme: as the time delays grow longer and the feedbacks more powerful, performance deteriorates markedly. (Numerous citations omitted.)

As in business, so in policy. In general, existing environmental statutes and regulations, based largely on the centralized command-and-control model, presuppose intervention in a simple system; that is, one in which the environmental, economic, and technological impacts of a specific intervention are easily determined. In reality, of course, virtually all of the systems at issue are complex systems. As Table 9.1 demonstrates, this has a number of critical policy implications. It is crucial to recognize, generalizing Senge's and Sterman's comments about business, that complex systems cannot be managed in the traditional sense. Perhaps, with enough understanding of the system and the wisdom to apply that understanding, their evolution can be guided.

This does not, however, mean that targeted intervention is always inappropriate. Rather, it means that targeted intervention will almost always have ripple effects, and usually unanticipated environmental and economic costs and impacts. In cases where the potential environmental impacts are large and difficult to reverse, it may be wise policy to impose such command-and-control requirements, even when the full systems implications are unknown. Moreover, such targeted requirements are probably appropriate where strong social values are involved. Most people, for example, would view releasing any significant amount of toxic metals into drinking water sources to be impermissible under any circumstances, making command-and-control an appropriate, perhaps even politically necessary, response. Beyond these interventions, however, effective policy will have to reflect the fact that both the global economy and the underlying natural systems, and their intricate, co-evolved linkages, are too complex for targeted command-and-control to be at all effective in moving toward sustainability.

This is especially true as it is highly likely that a more sustainable economy will be more, not less, complex than the current one, in that it will generate and contain more information. Intuitively, this follows from the fact that the failure of the present economy to achieve sustainability results from economic and environmental inefficiencies reflecting inadequate availability of information about effects of choices (externalities are not captured in the price structure), and a failure (or inability) to generate such information in the first place. For example, automobile oil is changed every 3,000 miles regardless of whether the chemical state of the oil, and the wear in the engine, justifies the change. Sensor systems monitoring the state of the oil and the engine, so that oil is changed only when it needs to be, would substitute increased information generation and integration into economic activity (in this case, changing oil) for unnecessary production of waste. Soil moisture sensors for agricultural operations, which would key demand for irrigation to actual soil conditions and thereby reduce water use to that actually required, would have a similar effect. In other words, a fundamental systems reason for environmental inefficiency is lack of appropriate feedback loops—and their creation throughout all technological systems as the global economy moves towards sustainability will make an already complex system yet more complex.

Table 9.1 summarizes these points. A particularly important consideration in the evolution of policy is the implicit assumption concerning information. The environmental approach of most countries, based on centralized command-and-control policy mechanisms, assumes that the regulators do, in fact, have all relevant information necessary to establish standards, an assumption that only holds if the system is very simple. At the heterogeneous stage, collaboration replaces mandate, and broader information inputs are obtained. The information collection process is, however, still under a central control. At the complex system stage, information is recognized to exist throughout the economic system, and control is largely replaced by metrics that measure the extent to which the economy is evolving towards sustainability.

Note, however, that targeted intervention may still be required. Two obvious examples are the global ban on CFCs, which cause stratospheric ozone depletion, and the ban on leaded gasoline in many countries. In both cases, the interventions have apparently been successful, and inordinate unexpected environmental or economic costs have not been identified. The point, therefore, is not that targeted interventions are inappropriate under all circumstances, but that they must be done cautiously and with the sure knowledge that the future path of the economy and technology, and its concomitant environmental impacts, will shift in response. Thus, for example, CFC cleaning systems in electronics manufacture were replaced in many cases with aqueous systems that required more energy consumption, and the ban on leaded gasoline reduced the compression ratio of automobile engines, thus reducing their engineering and economic efficiency and increasing gasoline consumption. In these instances, the costs attendant upon a targeted intervention probably were appropriate given the obvious and significant costs of not taking action; in many other cases, they are not. Accordingly, good public policy will never ban or significantly limit a material or technology without considering what the potential alternatives, and their impacts, are liable to be, and will do so only where the anticipated benefits are very large, so that not only the immediate and recognized costs, but the inevitable unanticipated costs, are outweighed.

More broadly, the measure of policy success—the means by which society determines whether progress is being achieved—shifts dramatically. Under the simple system model, policy success is simply compliance with the command-and-control requirements for remediation or end-of-pipe emissions. In system information terms, policy success is not linked to endpoint: Whether or not the targeted environmental impact is being mitigated, compliance by the regulated community is full satisfaction of social and legal obligations. Thus, for example, in a joint USEPA/Amoco study of a refinery, it was determined that existing regulations, while costly and stringent, included only a small percentage of actual benzene emissions. Because environmental impact and policy success measures were not linked, the refinery and EPA had both satisfied their legal obligations, without, however, achieving the desired environmental mitigation outcome (efficient reduction of benzene emissions).

As policy shifts toward the complex system model, however, compliance becomes less relevant. The measure of policy success becomes the degree to which measured change in metrics keyed to the response of both the economic and environmental systems occurs. Direct linkage between desired outcome (evolution towards sustainability) and policy initiatives builds a critical information feedback link, which enables policies to be adjusted based on new knowledge. Obviously, future environmental management, based on the complex system model, requires suitable sustainability metrics.

9.5 AN INFORMATION DENSE, SUSTAINABLE ECONOMY

9.5.1 The Automotive Technology System, Information, and Complexity

The discussion in Chapter 4 about the automotive technology system as a case study raises another point regarding an economically and environmentally efficient, sustainable economy. As noted in that discussion, the modern automobile has substantially improved its environmental efficiency while maintaining performance levels; it has, at the same time, become a safer and more desirable product (air bags, better audio systems, better braking systems). How was this achieved, and what might this analogy say about the information density and complexity of a whole economy that is also becoming more environmentally and economically efficient?

The performance of the modern automobile reflects a number of incremental improvements: reductions in vehicle weight, better aerodynamic design, reductions in tire rolling resistance, new catalytic engine systems, more efficient engines and drivetrains. But there is one common theme underlying the evolution of the modern automobile and the infrastructure within which it functions: It has become a much more complex system with a far higher information content than its predecessors (Figure 9.1), and it is increasingly linked to its external environment. It is becoming a subsystem tightly linked into a yet more complex automotive transportation system.

Whereas older cars had minimal electronics, newer ones have substantial computing systems; in fact, a typical modern automobile has more computer-processing power than the first lunar lander had in 1969. These systems not only enhance the efficiency of individual systems, but create new information linkages within the car itself, making it a more integrated system as well. Older cars were linked together, where necessary, mechanically; newer vehicles are linked by systems of sensors feeding into multiple computers. These chips not only link systems within the car, but link the car much more tightly with external environmental conditions, such as air density. Increased efficiency comes from more information, not less. Computers made the energy and environmentally efficient car possible in part by more tightly integrating previously disparate subsystems (by creating more feedback loops).

Producing a more complex and efficient artifact requires, in turn, more sophisticated design tools and manufacturing technologies. For example, "lightweighting" a car —designing it to weigh less for better fuel economy while providing equivalent performance and safety—requires precision manufacturing and more complex manipulation of non-traditional materials such as aluminum—it is a far more information intensive activity. Germany's Audi company, for example, believes that only the advent of supercomputing power provided them with the necessary capability to design complicated lightweight components that, in turn, are an important contributor to a lighter vehicle.

A similar evolution toward greater complexity marks the car's relationship to its infrastructure. With old cars, virtually the only information link between the automobile and the external environment was the driver; today, however, sensor systems not only monitor the oxygen content of air flows and exhaust systems, but road conditions

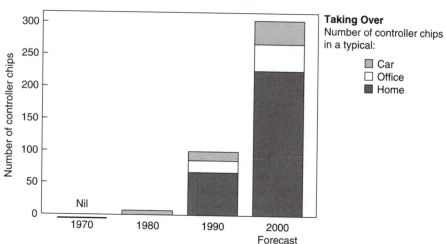

**Substitution of Information Management
for Inefficient Resource Use**

FIGURE 9.1   Substitution of information systems for inefficient resource use in the automobile, home and office. These trends are, if anything, accelerating, and may well have profound implications for resource consumption patterns in the future.

Source: Based on *The Economist*, Survey of Computer Industry. September 17–23, 1994, p. 20. based on Motorola data.

as well. Newer systems already deployed in leading countries such as Japan even map the car's geographical position, provide information on current road conditions, optimize real-time routing options, and pay tolls electronically without the need to stop. The technologies that will permit ongoing communication between road networks and automobiles—in essence integrating the automotive built infrastructure, the automobile, and the driver into one automotive transportation system, which can then be optimized for real-time efficiency by, for example, use-sensitive automatic roadway pricing technologies—already exist. Mercedes-Benz, for example, has designed and built a car that is fully integrated into the World Wide Web: the car as information appliance, just another node on the Internet. The next step, integrating different modes of transportation for people and cargo, is already beginning.

Thus, the automotive technology system example provides powerful support for the hypothesis that a more economically and environmentally efficient economy will be more complex in a number of ways. Individual systems, such as the automobile, will become more complex: They will sense, transmit, and respond to more information in providing their function. At a higher level of function, previously weakly linked systems, such as the automobile and the ambient air conditions and the automobile and its infrastructure, will become far more strongly linked into new metasystems, which in turn will both be optimized and linked into higher level systems as well. A sustainable economy will not be the agrarian vision that many now have; rather, it will be a more engineered, more complex, more tightly linked system. It is doubtful that even developed societies yet have the scientific and technical knowledge to know what that means, or the social and political capability to act upon it reasonably.

9.5.2 Defining the Information Density of an Economy

One way of thinking about a more complex economy is in terms of its information density, especially if greater information density is likely to be a necessary, if not sufficient, requirement for greater economic and environmental efficiency. It is, of course, also apparent that a more complex economy, in itself, does not guarantee such efficiency: One could simply devise more numerous and more complex ways of producing more onerous environmental impacts.

A very rough estimate of the information density might be obtained by using economic data alone, using summed data on consumption in "information sectors" such as books, cable television, telephone, CDs, and comparing the ratio of that sum to economic activity as a whole. This approach is appealing because the data are available, but it could yield seriously misleading results because of the variable cost trends in information technology, and the fact that much of the value added of information products comes not from their information content (which is what one wants to measure), but mark-ups through the supply chain. Thus, for example, it costs about $1.50 to manufacture a CD that sells for about $15.00; the $13.50 difference represents costs associated with packaging, transportation, overhead on stores, and profit for various firms involved in distribution and retailing.

The difficulties in using this approach are illustrated in Table 9.2, which uses U.S. Bureau of Economic Analysis data to track information industry economic performance as percent of GDP for selected years from 1980 to 1994 for the United States. Sectors selected include Manufacturing, Durable Goods, Electric and Electronic Equipment; Communications (including Telephone and Telegraph and Radio and Television Broadcasting); Motion Pictures; Amusement and Recreation Services; and Printing and Publishing. Of these, electronics manufacturing and communications are the dominant categories, with publishing third, so if dollars of activity represented a viable approach, one would expect to see, in line with the data on information sector activity previously presented, a significant increase in the percent of GDP represented by these activities. This does not happen: Over a 15-year period the increase is from

TABLE 9.2   Traditional Information Sector Share of GDP (Billions of 1996 Dollars, Rounded)

Year	1980	1987	1988	1989	1990	1991	1992	1993	1994
GDP	2708	4692	5050	5439	5744	5917	6244	6550	6931
Sector									
Electronics Mfg.	55	83	88	97	95	98	99	112	130
Communications	69	125	132	136	147	154	161	173	188
Motion picture	6	15	16	20	20	20	20	22	25
Amusement & Recreation	14	27	30	34	39	42	48	49	52
Printing & Publishing	33	62	69	72	74	76	79	82	86
Sectors Total	177	312	335	359	375	390	407	438	48
As Percent of GDP	6.53	6.65	6.63	6.6	6.53	6.6	6.52	6.69	6.94

Source: Statistical Abstract of The United States, 1996, Tables 685 and 686; US Bureau of Economic Analysis data at www.bea.doc.gov/bea/gpxind-d.htm++curr

about 6.5 percent to about 6.9 percent. Either the increase in information capacity in the U.S. economy is illusory, or these data do not capture the trend. The latter appears most likely. Accordingly, a more rigorous approach focused on actual information content would be highly preferable.

Conceptually, the information density of an economy is somewhat analogous to physical density, and thus can be given by a formula based on that for physical density (volume divided by mass):

$$D_i = \frac{V_i}{E_a}$$

where D_i is the information density of the economy in bits per dollar, V_i is the volume of information in the economy in bits, and E_a is the economic activity in the economy measured in dollars. Unlike physical density measures, however, information density involves a time domain because both stocks and flows of information are involved. Thus, information density might have to be an averaged figure (perhaps over a year to match GDP data). Economic activity is relatively easy to measure (although even here appropriate caution must be exercised: aspects of productivity, service sector activity, and enhanced quality of life due to improvements of existing artifacts are not well understood). Quantifying the volume of information in the economy is the problematic step.

Conceptually, one way of defining V_i is as the number of bits (or bit equivalents, for analog systems) communicated through the communications networks of the economy or consumed in a given period, including data, voice, and video (volume telecommunicated, or V_t); the number of bits consumed in the economy through, for example, listening to music, or watching a video tape (volume consumed, or V_c); the number of bits generated within artifacts, such as cars, airplanes, and coffeemakers (volume in artifacts, or V_a); the number of bits generated within facilities and infrastructure, such as manufacturing facilities, administrative buildings, retail outlets, fast food establishments, and the like as part of their operations (volume in facilities, or V_f); the number of bits published in other than electronic media, and not duplicated there (V_p); and the information content in all other, residual uses (V_r).

Some of these terms, such as the volume telecommunicated, can be derived from existing data that are already collected. Based on the observation that information density should probably reflect the use of information, rather than simply the capability to produce information (e.g., a silicon wafer in itself does not contribute to the information density of the economy), other terms might be estimated by multiplying the information capability of the relevant universe by the number of times that capability is accessed. Thus, an estimate for bits consumed could be derived by multiplying the number of bits stored in various media by the times the media unit is accessed, and multiplying that figure in turn by the audience per time accessed. The number of bits generated within artifacts also might be relatively easy to estimate based on the capability of the chips embedded in the devices and the access rate. A similar process, adjusted for double counting, might be applied to bits generated within facilities and infrastructure. A figure for bits published in non-electronic media forms should be relatively easy to estimate, although double-counting might be a problem here as well. A little practice would be useful in determining how easily rough figures for these factors can be derived, and whether some are so much more important than others that, for all practical purposes, they drive the analysis.

Thus,

$$V_i = V_t + V_c + V_a + V_f + V_p + V_r$$

Defining these amounts will be difficult, especially as the terms are not orthogonal; some of the information consumed, for example, is previously transmitted.

The easiest way to measure E_a is in terms of dollars of economic activity, but this has some measurement difficulties as well. Many of these, such as costing "noneconomic" but productive work such as housework or raising children, are familiar to economists, however, and reasonable valuation methods can be used to impute proper figures. More fundamentally, in keeping with the industrial ecology approach, which encompasses both economic activity and associated externalities, E_a should be considered as the sum of measured economic activity, or E_m, and externalities, or E_x:

$$E_a = E_m + E_x$$

The difficulties of quantifying externalities are substantial, but not impossible if absolute precision is not required.

This provides the full equation:

$$D_i = \frac{V_t + V_c + V_a + V_f + V_p + V_r}{E_m + E_x}$$

Estimating this indicator for the period from the beginning of the Industrial Revolution until the present could be an interesting way of determining whether the vaunted "information revolution" is, in fact, occurring. It might also be one indicator, albeit insufficient without others, of progress toward a more environmentally and economically efficient economy. Much more research is required in this area to achieve an understanding of the information economy, trends in information density, and the relationship between those trends and increasing sustainability. This approach also raises interesting questions in related areas. For example, an interesting question for further research is whether V_i can be broadly defined in such a way as to generate a quantitative measure of the complexity of society as a whole, rather than just the economy.

REFERENCES

Adriaanse, A. *Environmental Policy Performance Indicators: A Study on the Development of Indicators for Environmental Policy in The Netherlands.* The Netherlands, Ministry of Housing, Physical Planning and Environment: 1993.

Costanza, R., L. Wainger, C. Folke, and K. Maler. "Modeling complex ecological economic systems." *Bioscience* 43(8): 545–555 (1993).

Graedel, T.E. and B.R. Allenby. *Industrial Ecology and the Automobile.* Upper Saddle River, NJ, Prentice-Hall: 1997.

Hammond, A., A. Adriaanse, E. Rodenburg, D. Bryant, and R. Woodward. *Environmental Indicators: A Systematic Approach to Measuring and Reporting on Environmental Policy Performance in the Context of Sustainable Development.* Washington, DC, World Resources Institute: 1995.

Lawton, J.H. "Ecological experiments with model systems." *Science* 269: 328–331 (1995).

Luhmann, N. *Ecological Communications.* (J. Bednarz, Jr., trans.) Chicago, University of Chicago Press: 1989.

Root, T.L. and S.H. Schneider. "Ecology and climate: research strategies and implications." *Science* 269: 334–341 (1995).

Senge, P.M. and J.D. Sterman. "Systems Thinking and Organizational Learning: Acting Locally and Thinking Globally in the Organization of the Future," Sloan School of Management, MIT, Cambridge, MA.

EXERCISES

1. One of the most famous technological contests in the literature is that between two videotape standards, Betamax and VHS. Although Betamax was considered by many to be the superior standard, VHS grew to dominate the market. Explain this result in terms of complex system dynamics.

2. Consider two regulations, one of which bans a toxic material and the other of which imposes no direct restrictions on it, but requires that any release of it, either unintentional or as part of the life cycle of a product, be reported publicly.

   a. Which is most likely to have unintended consequences, and what might some of these consequences be?

   b. Which is most likely to be economically inefficient?

   c. What are the likely effects on the behavior of the regulated community from each regulation?

   d. Which would make you feel more secure if you lived next to a factory using the material?

3. In section 9.5.2, a formula for the information density of an economic system is derived. It is a viable hypothesis that not just economic activity, but society itself, will get more complex over time as increasing economic and environmental efficiency is achieved. Derive and justify a simple equation for the information density of a society.

C H A P T E R   1 0

Risks, Costs, and Benefits

10.1 OVERVIEW

Policy is ineffectual, even harmful, unless it bears a reasonable resemblance to the underlying phenomena it is intended to address. A significant reason for the inability of many governments to implement environmental regulation in a manner that is both economically and environmentally efficient is the lack of any integrated mechanism to understand environmental risk in a comprehensive way, and to link this understanding with other risks, especially economic, social, and technological, which any choice inevitably entails.

Traditionally, separate analytical toolboxes have been used to evaluate economic risk and environmental risk. Economic risk is frequently evaluated using cost/benefit analysis methodologies, which attempt to quantify expected and frequently probabilistic costs and benefits of a course of action. Environmental risk is evaluated using risk assessment techniques, the application of which in practice has been limited primarily to human impacts, and, even more specifically, to human cancer impacts. Cost/benefit analyses tend to be inclusive, at least to the extent that issues can be quantified; environmental risk assessments tend to be limited and reductionist, reflecting the specificity of the defined endpoints. Environmental Impact Assessments, or EIAs, required in Europe for certain projects, and the similar Environmental Impact Statements, or EISs, required for certain Federal projects in the United States under the National Environmental Policy Act, are broad environmental assessments that, in scope, are more similar to cost/benefit analyses than environmental risk assessments.

This chapter provides a brief conceptual introduction to these methodologies. Each methodology taken alone, however, is inadequate to inform systems-based, industrial ecology policy making. Accordingly, the chapter ends with a suggestion for developing and implementing Comprehensive Risk Assessment tools, or CRAs, which can begin to present integrated assessments of environmental risks that can then be evaluated along with other costs and benefits associated with policies and decisions with substantial environmental dimensions. Such integrated decision methodologies should be developed for use at any level of the systems hierarchy—sustainability, industrial ecology, industrial ecology infrastructure, and applications to practice—by any competent entity. In this way, CRAs linked with metrics, which effectively capture progress toward economic and environmental efficiency at all system levels, form a necessary competency for approaching sustainability. Although these capabilities are currently far beyond the state of the art, encouraging the development and diffusion of such practices should be a major policy goal, because this process in itself will also educate people to take systems-based, comprehensive approaches to environmental concerns.

10.2   COST/BENEFIT ANALYSIS

A common engineering and economic evaluation tool for evaluating potential courses of action is a *cost/benefit analysis*, in which the costs and benefits of the activity are quantified and then compared. Such analyses can be relatively formal and technical or relatively informal, depending on the context and purpose of the exercise. Because cost/benefit analyses are usually quantitative, uncertainties or unquantifiable issues, such as moral or ethical questions, may be difficult to include in such a procedure. Ideally, these issues are integrated into the decision-making process at some point; in practice, they may be discounted or ignored completely.

Cost/benefit analyses can be carried out at many different levels: at a project or firm level (e.g., the business plan process), at a national level, or even at an international level. Regardless of the scale, it is useful for the industrial ecologist to consider several questions when evaluating such a study:

- What assumptions, implicit and explicit, have been made? Are they technically appropriate? Are they transparent? For example, a common, sometimes deliberate, and usually invalidating assumption is that technology is static under conditions of economic change or regulatory intervention, which simplifies an analysis, but is almost always never true, and almost always overstates costs.

- Who is performing the analysis, and what are their interests in the outcome? Has the analysis been performed solely to support a predetermined position?

- What are the geographical and temporal distributions of the costs and benefits? Are certain geographic, economic, or racial groups receiving a different set of costs and benefits than others? Is the cost/benefit study performed from the perspective of a group potentially receiving disproportionate gains or losses? Are costs being exported from one nation to another?

- What, if any, discount rate is being applied to future costs to justify present benefits? Are there unrealistic assumptions built into the model (e.g., revenue streams are all invested for the future, rather than consumed in the present)?

- Are there any ethical or value considerations, or other unquantifiable dimensions, that have not been included in the analysis? To what extent are assumptions involving values integral to the assessment, and are all such assumptions transparent and justifiable?

- What are the significant uncertainties related to the analysis, and can they be quantified? What is the sensitivity of the cost/benefit analysis to the elements identified in the preceding questions?

The point of such an assessment is not to invalidate a cost/benefit assessment (unless, of course, it is performed so poorly as to be useless). Rather, it is to create a context within which the information derived from the analysis can be usefully integrated into decision-making or policy formulation.

10.3   RISK ASSESSMENT

Risk assessment may be defined as a method for evaluating risks so that they may be reduced, prioritized, avoided, communicated, or managed. "Risk," in turn, may be defined as the probability of suffering harm from a hazard. While professionals tend to think of risk as objective, risks are regarded by most people as intensely subjective. Thus, studies that have asked experts and informed lay people to rank the same activities have consistently demonstrated wide variances between the groups, as shown in Table 10.1. The same pattern is true for environmental issues: Table 10.2 shows the comparative rankings of such issues provided by the public and U.S. Environmental Protection Agency experts. This difference in perception appears to arise because the public integrates a number of subjective factors into their determination of risk, particularly the following:

- The extent to which the risk appears to be controllable by the population at risk (note in Table 10.1 that college students, who ride bicycles a lot, and thus feel that they are in control on them, rank that risk much lower than the experts).
- Whether the risk is feared, or even dreaded (the focus of much risk assessment on human cancer arises, in part, because of the dreaded nature of the disease).
- The extent to which the risk is imposed rather than voluntarily assumed (note from Table 10.1 that the League of Women Voters ranks contraceptives as much less risky than the experts, probably reflecting not only voluntary assumption of risk, but also familiarity with contraceptives, and quite possibly a personal understanding of the trade-offs in risk involved in not using contraceptives as well).
- The extent to which the risk is easily observable, especially by the at-risk population, and, if observable, is manageable by science (nuclear power issues, where the risk is neither readily observable nor easily understood, and where the possible endpoint of exposure, cancer, is particularly dreaded, stand out in Table 10.1).
- Whether the victims are especially sympathetic and vulnerable (particularly children).
- The extent to which the risk is new and unfamiliar, and unquantifiable or previously unknown to science.
- The extent to which the victims are identifiable as individuals, as opposed to statistical groupings (media coverage of air crashes, for example, tends to focus on individual victims, particularly children, which may explain why the public perception of the risk associated with them differs significantly from the expert assessment).

TABLE 10.1   Perception of risk from most risky (1) to least risky (30) by three target audiences: educated and politically involved female citizens; college students; and experts. Some reasons for these varied perceptions are provided in the text.

Activity or technology	League of Women Voters	College students	Experts
Nuclear power	1	1	20
Motor vehicles	2	5	1
Handguns	3	2	4
Smoking	4	3	2
Motorcycles	5	6	6
Alcoholic beverages	6	7	3
General (private) aviation	7	15	12
Police work	8	8	17
Pesticides	9	4	8
Surgery	10	11	5
Fire fighting	11	10	18
Large construction	12	14	13
Hunting	13	18	23
Spray cans	14	13	26
Mountain climbing	15	22	29
Bicycles	16	24	15
Commercial aviation	17	16	26
Electric power (non-nuclear)	18	19	9
Swimming	19	30	10
Contraceptives	20	9	11
Skiing	21	25	30
X-rays	22	17	7
High school and college football	23	26	27
Railroads	24	23	19
Food preservatives	25	12	14
Food coloring	26	20	21
Power mowers	27	28	28
Prescription antibiotics	28	21	24
Home applicances	29	27	22
Vaccinations	30	29	25

From Slovic, P. Perception of Risk, *Science* 236: 280–285.

- The extent and type of media attention (sensationalist as opposed to factual reporting).
- The perceived equity of the distribution of the risk among different groups.

This last point is particularly important, as it is a significant differentiator between expert approaches to risk assessment and public perception of risk. If the public believes that, to some extent, a risk has been assumed by the at-risk population, it is more likely to accept the risk as legitimate (this is a major reason, for example, why

Table 10.2   COMPARATIVE RANKINGS OF ENVIRONMENTAL PROBLEMS

Public Perception of Risk (Priority Ranking)		Corresponding U.S. EPA Ranking (*Unfinished Business*, 1987)	
1.	Chemical waste disposal	16.	Hazardous waste sites—active
		17.	Hazardous waste sites—inactive
2.	Water pollution	9.	Direct point source discharges
		10.	Indirect point source discharges
		11.	Nonpoint source discharges
3.	Chemical plant accidents	21.	Accidental releases—toxics
4.	Air pollution	1.	Criteria air pollutants
		2.	Hazardous air pollutants
5.	Oil tanker spillage	22.	Accidental releases—oil
6.	Exposure on the job	31.	Worker exposure
7.	Pesticide residues on food	25.	Pesiticide residues on food
8.	Pesticides in farming	26.	Application of pesticides
		27.	Other pesticide risks
9.	Drinking water	15.	Drinking water
10.	Indoor air pollution	5.	Indoor air pollution
		30.	Consumer product exposure
11.	Biotechnology	29.	Biotechnology
12.	Strip mining	20.	Mining waste
13.	Nonnuclear radiation	6.	Radiation other than radon
14.	The greenhouse effect	8.	Global warming

The following U.S. EPA problem areas were not ranked in the public poll.
3. Other air pollutants; 4. indoor radon; 7. stratosphereic ozone depletion; 12. contaminated sludge; 13. estuaries, surface waters, and oceans; 14. wetlands; 18. nonhazardouse waste sites— municipal; 19. nonhazardouse waste sites—industrial; 23. releases from storage tanks; 24. other groundwater contamination; and 28. new toxic chemicals

individual smokers have generally failed to win jury trials against tobacco companies in the United States). Thus, for example, the public is more likely to be concerned when the costs, benefits, and risk of a particular activity are disproportionately allocated among groups ("distributed justice", in other words) than the expert risk assessor, interested primarily in cumulative increases or decreases in absolute risk. The reader will, of course, immediately recognize the close analogy between these dimensions of risk and neoclassical economic theory, where the emphasis is on wealth formation at micro and macro levels, with little attention paid to distribution of that wealth (except inasmuch as distribution patterns affect wealth generation).

Risk assessments may be highly qualitative, such as those we all make in the course of daily living ("Should I drive to work on a crowded highway?"). The type of risk assessment commonly used in environmental regulation, however, is highly quantitative. All risk assessments deal with uncertainty and rely on assumptions that may or may not be transparent and appropriate to the circumstances. If done well, they may be regarded as best estimates; if done poorly, they can lead to inappropriate choices, inefficient use of scarce resources or, worse yet, to increases in actual risk.

Risk assessments, at least the more rigorous ones, generally consist of five stages (technical information on risk assessment methodologies can be found in the references listed at the end of this chapter):

1. The *hazard identification stage* is where one determines whether a hazard actually exists. For a number of reasons, such as the need to extrapolate from uncertain data, this may not be a straightforward determination.

2. The second stage is to determine the *delivered dose*. This step may involve detailed knowledge of the environment within which the release occurs, the chemical form of the dose as actually delivered, and the physical system within which the materials under consideration are found. It is incorrect, for example, to take the total lead burden of a water column from a stream as the actual dose: lead adsorbs onto sediment easily, and is not readily bioavailable in that form.

3. The third stage is to assess the *probability of an undesirable effect as a result of the delivered dose.* Among the major uncertainties at this stage are the models that extrapolate from high, almost fatal doses of toxicants in one species to very low exposures in other species.

4. The fourth stage is to *determine the exposed population.* The total risk impact is the number of individuals exposed times the probability that the indicated dose will cause the undesirable effect. This may be expressed mathematically as:

$$I = NP\,(d)$$

where I is the total risk impact, N is the number of individuals exposed, and P(d) is the probability, P, that the indicated dose, d, will cause the effect.

5. The final stage of the assessment is to *characterize the risk.* This involves explaining quantitatively the risk posed by the risk agent, and ideally requires that assumptions, uncertainties, and other potential confounding factors be identified.

In most cases, the risk will have been identified and assessed as a part of a larger project or program. The follow-up to risk assessment, therefore, is usually to integrate the risk information into the broader policy activity and communicate it to various stakeholders in the policy process ("risk communication"). In this step, the subjective and quantitative are both involved, and there should always be an assessment of the way the results are presented because of the opportunity for communicating deliberate or unintentional misrepresentation of the risks. For example, it is well known that air travel is safer than car travel—or is it? As Table 10.3, based on British data, shows, this is true on a per kilometer basis but not on a per journey basis, and on a per hour basis, the two are equivalent. Motorbikes are far more risky than walking on a per journey basis, but the difference between them is less than two to one on a per kilometer basis. Buses are less risky than any other mode of transportation on either a per journey or per hour basis, and virtually as safe as air travel on a per kilometer basis (unfortunately, there is no bus service between London and New York). The lesson

TABLE 10.3   Relative Risk Comparisons (British Fatality Rates, 1992 Data) Even using the same base dates, how risk is expressed can significantly change the impression created (compare air with car traffic on a per journey, per hour and per kilometer basis).

Mode of transport	Per 100m passenger:		
	journeys	hours	km
Motorcycle	100	300	9.7
Air	55	15	0.03
Bicycle	12	60	4.3
Foot	5.1	20	5.3
Car	4.5	15	0.4
Van	2.7	6.6	0.2
Rail	2.7	4.8	0.1
Bus or Coach	0.3	0.1	0.04

Based on: *The Economist*, January 11, 1997, p.57

here is that risk is difficult both to define and communicate, and caution is advisable in interpreting even quantitative data.

The final step following the assessments, studies, and communication of the results is to develop and implement mechanisms to reduce or control the risk in light of other factors, such as statutory requirements or economic conditions. This generally falls under the rubric of "risk management."

10.4   COMPREHENSIVE RISK ASSESSMENTS (CRAS)

The field of risk assessment as currently practiced by environmental specialists is analogous to the pre-industrial ecology state of environmental science and regulation. As currently practiced, risk assessment tends to deal with specific materials or localized conditions and ad hoc, limited endpoints (Table 1.1). In practice, most environmental risk assessments are performed for regulatory purposes, and most focus almost exclusively on human endpoints, and, even there, mostly on carcinogenesis. Such limited studies are, indeed, necessary, but are not in themselves sufficient to support intelligent public policy choices. Important dimensions of risk in the real world that tend not to be considered in such specific risk assessments include:

- Trade-offs among environmental impacts, or environmental and other impacts (e.g., economic, employment, quality-of-life).
- The need to address multiple endpoints rather than the most feared or the one where most data can be gathered (e.g., carcinogenesis).
- The inherent ambiguity of complex environmental and economic systems within which activities giving rise to risk occur.
- The need to include not just the specific activity under consideration but also alternatives and options in the risk assessment.

- The very difficult but critical role played by different values assigned by different stakeholders to the same economic or environmental factor.

It is not that risk assessment as currently practiced is necessarily improper or inadequate when properly scoped or not valuable in informing remediation and compliance programs. But to support the implementation of industrial ecology and the process of approaching sustainability, a broader concept of risk assessment that builds on, but is not limited to, current practices is necessary. This is the *Comprehensive Risk Assessment (CRA)*.

There are a number of dimensions of risk that a CRA should comprehend:

- A CRA should support the capability to make trade-offs among environmental impacts, and between environmental impacts and other socially desirable goals such as economic efficiency, employment, cultural values, or quality-of-life. Although opportunities to achieve unequivocal environmental benefits with no associated costs do exist, they are relatively trivial; the more difficult cases are those where trade-offs are inherent in the problem. It is here that policy making requires CRA support.

  An example may make this clearer. Consider the new class of materials known as high temperature superconductors. Once integrated into energy production and consumption technologies, they promise enormous energy savings. On the other hand, it is highly probable that they will embody toxic materials such as copper, mercury, or thallium. Traditional risk assessment methodologies, which focus on toxics, would tend to support a policy of avoiding the use of toxic materials and thus eschewing superconductors. When the potential substitution effects are considered, however, and the system is considered as a whole to include the impacts of energy production, the hypothesis that superconductors should be widely diffused throughout the economy because of the environmental benefits of greatly reduced energy demand becomes much more tenable. Without some form of a CRA methodology, the difficult trade-offs involved in this technology cannot be responsibly evaluated.

- CRAs should address multiple endpoints in risk assessment, rather than that endpoint that is most feared or where the most data can be gathered. Moreover, the endpoints should not be related only to human health and environmental impact, but should reflect all aspects of the situation that might cause diminution in quality-of-life. Where uncontroversial quantification cannot be achieved, quasi-numerical methods to integrate multiple endpoints can be used.

- The complex environmental and economic systems with which CRAs must deal are highly uncertain for a number of reasons, such as data unavailability, high levels of uncertainty associated with the data that are available, and the arguably deterministically chaotic nature of the systems themselves. Accordingly, rigorous quantitative analyses resulting in specific values should not be the only tools that are accepted. The false security gained by acting based on the quantification of a small part of a complex system is paid for in suboptimal responses of the system as a whole: Complexity and difficult ethical questions do not just disappear when they are avoided. In this sense, limited risk assessments are examples of what Daly and Cobb have called the "fallacy of misplaced concreteness": Substitution of partial quantitative analyses for a more full—and more valid—comprehensive assessment.

An important associated issue is the failure of most risk assessments to evaluate risks across the relevant life cycle. For example, a typical risk assessment of lead solder would demonstrate unequivocally that lead is a toxic material and should thus be phased out. On the other hand, doing a broader risk assessment that looks at the life cycle of lead solder compared to the life cycle impacts of some alternatives such as indium or bismuth solders results in a much more ambiguous finding. Importantly, the critical environmental impacts associated with indium and bismuth are all in the mining and processing life cycle stage, and are not a result of their inherent toxicity. Bismuth, for example, is a byproduct of lead mining, so increasing demand for bismuth by switching away from lead solder might actually increase, rather than decrease, lead mining activity. Thus, even a standard risk assessment of these materials, based on their toxicity but not considering the life cycle of the materials, would have missed the relevant environmental risks.

- A CRA must include evaluations of the appropriate alternatives and options. In the global economy few regulatory restrictions are imposed, for example, without generating the use of alternatives, and few courses are taken that do not involve ancillary impacts. The risk associated with these options and ancillary impacts are properly considered as part of the original decision that led to them.

- An issue that has bedeviled traditional risk assessment, and is even more difficult in the context of a CRA methodology, is the difference in perception of risks and their relative priority depending on fundamental values. Risk assessment attempts objectivity; risk perception, upon which action is usually implicitly or explicitly based, is strongly subjective. This is not wrong, as experts sometimes argue, nor will it go away with further education of the public (although it may well be more informed). In risk, as in other areas, people believe in justice and involve their values in their assessments: This is simply a fact that must be recognized in developing any CRA methodology.

It is unquestionably difficult to quantify values, especially regarding environmental issues where fundamental disagreements often arise. Some religions, for example, see family planning as against the will of God, while other groups believe that a failure to control human population is immoral. Some in the American West believe that any attempt to limit their use of public lands, which constitute a large area in that region of the country, are an abridgment of their fundamental freedom, while others believe that a failure to control such exploitation is an unacceptable squandering of the heritage of future generations. In fact, the difficulties posed by differing values are often used to claim that comprehensive risk assessment methodologies are virtually impossible.

There are, however, a number of ways (albeit somewhat controversial) to quantify material containing a significant value dimension. For example, the Swedish EPS

(Environmental Priority Strategies for product design) system develops Environmental Load Units (ELUs) using a "willingness to pay" values quantification process, in which surveys indicate what people would be willing to pay for the environmental benefit at issue.

Moreover, especially in countries such as the United States where different stakeholders are both polarized and vocal, it is possible to overemphasize the differences in values. In this regard, the survey of environmental attitudes in the United States by Kempton et al. is quite interesting. For example, data indicate the existence of a fairly strongly felt moral obligation to other species among Americans (figures are percent agreeing with the question; the sawmill workers are from the Pacific Northwest, where their jobs are threatened by the spotted owl):

Question 16. Justice is not just for human beings. We need to be as fair to plants and animals as we are to people.

Earth First!	Sierra Club	Public	Dry Cleaners	Sawmill Workers
97	85	90	83	63

Question 50. Other species have as much right to be on this earth as we do. Just because we are smarter than other animals doesn't make us better.

Earth First!	Sierra Club	Public	Dry Cleaners	Sawmill Workers
97	78	83	83	56

Only when the most difficult values trade-offs are posed—such as human life versus species extinction—does the broad consensus fray, and even here Kempton et al. note (111–112) that, "still 40 to 50 percent of the moderate three subgroups would rather that humans suffer or die than cause extinction of 'an entire species' . . . we find this surprisingly high."

Question 84. I would rather see a few humans suffer or even be killed than to see human environmental damage cause an entire species go [sic] extinct.

Earth First!	Sierra Club	Public	Dry Cleaners	Sawmill Workers
90	48	43	56	22

10.5 DEVELOPMENT OF A COMPREHENSIVE RISK ASSESSMENT METHODOLOGY

Given the difficult analytical and subjective issues raised by risk assessment, there has in the past been some skepticism about the possibility of developing a CRA, or, indeed, about the value of risk assessment itself. As the preceding discussion indicates, however, regulatory decisions and policy development should be based on the best available information, even if it is incomplete or uncertain, and risk assessments, properly reflecting system attributes and scope, are an important linchpin for these processes. The challenge is to create CRA methodologies that are quantitative enough to be integrated into broader cost/benefit assessments, and yet at the same time transparent enough so that assumptions about values, subjective perceptions, and uncertainties are apparent.

Accordingly, in this section a generalized conceptual CRA model, based on the recognition that there are qualitatively different categories of risk associated with environmental concerns, is presented. The model uses a risk taxonomy based on that adopted by the government of the Netherlands (see Table 10.4), which establishes three categories of risk. The first category of risks are those concerning damage to biological systems in general and humans in particular. The second category of risks are those that aesthetically degrade the environment, but may or may not damage biological systems. The final category of risks are those involving damage to fundamental planetary systems.

This risk categorization—and it is only one of many possible—can be used to derive an illustrative CRA methodology. To do so, first consider the generic risk equation used to compute the risk of the effect of a particular dose on a given population:

$$I = NP(d_i)$$

where I is the risk, N is the number of individuals at risk, P is the probability of an effect, and d_i is the level of dose of the *i*th source of impact. Using this simple equation, we can then use the Dutch risk categories to develop a conceptual semi-quantitative model.

For the first category, damage to biological systems, the equation can simply be written as:

$$B = bNP(d_i)$$

where B is the comprehensive biological risk and b is a weighting factor, agreed to by social consensus, that reflects both the objective and subjective value placed on biolog-

---

TABLE 10.4   Risk Categories

Damage to Biological Systems	Esthetic Degradation	Damage to Planetary Systems
Acid rain (acquatic effects)	Acid rain (material corrosion)	Biodiversity loss
Air toxics (including smog)	Oil spills (visual effects)	Changes in ocean circulation
Groundwater degradation	Loss of visibility	Global warming
Hazardous waste sites	Loss of opportunity for wilderness experiences	Ozone depletion
Herbicides, pesticides		Loss of arable land
Oil spills (wildlife effects)		Degradation of global water resources
Surface water degradation		
Radionuclides		
Toxics in sediments		
Toxics in sludge		
Loss of habitat		

ical systems, including human systems, by society. If appropriate given cultural values, this term can be broken into two terms reflecting different weighting for human effects and effects on other biological systems.

Risk associated with aesthetic degradation can be expressed in a similar manner as follows:

$$A = aNP(d_i)$$

where A is the aesthetic risk, N is the number of people affected by aesthetic degradation (including those who may not be physically present, but who value the impacted environment), P is the probability of an effect for the dose d of the ith source of impact, and a is again a weighting factor reflecting societal consensus. It is likely that "a" would be less than "b," reflecting the fact that aesthetic degradation is felt by most people not to be as serious in itself as damage to biological or human systems.

The final category of environmental risk, damage to planetary systems, may be similarly determined. Weighting factors should be high because the effects in this category impact the stability of systems involving the entire planet. It is also necessary to integrate global effects over time because they may extend for several generations to many centuries:

$$G = g\int_{t_0}^{t_1} N(t)\, P(d_{i,t})dt$$

where G is the global risk, g is the weighting factor, and the integration is performed from the present time (t_0) through the lifetime (t_1) of the substance or insult in question. Notice that the dose and the affected population are time-dependent.

Given these three equations, it is possible to develop a simple CRA:

$$CRA = B + A + G$$

where the comprehensive risk equals the sum of the biological (B), aesthetic (A), and global (G) effects for any particular subject of assessment.

It is worth emphasizing that this is a general, conceptual model for treating risk in a relatively comprehensive way. Although it includes values in the factors, for example, it does not negate the difficulty of reconciling differing values, nor is it meant to. Moreover, no risk assessment should be reported without providing the context necessary to evaluate the universe of what may be called "concomitant risk"—what would the result be if the option under consideration were not implemented? What alternatives are implied, and what are their risks? This implies the need for a broader context for individual CRAs, and their integration into cost/benefit analyses.

10.6 INTEGRATING RISK, COSTS, AND BENEFITS INTO A COMPREHENSIVE POLICY SUPPORT ASSESSMENT

The final step in developing a more rigorous basis for regulatory decisions or policy formulation is the integration of CRAs and cost/benefit analyses into a final product which for convenience can be called a *Comprehensive Policy Support Assessment* (*CPSA*). A conceptual four-stage methodology for doing so can be outlined:

1. *Option identification* is the first stage. Whether it is regulatory action directed at a single material, imposing an emissions technology, or simply the initial process of identifying a problem and seeking potential solutions, it is important from the beginning to identify likely consequences of each course of action. Identifying the full set of options will usually require participation of a number of stakeholders, so an open and collaborative regulatory and policy development system is required.

2. CRAs should be *developed for each option*. For example, if the concern is emissions of chlorinated solvents from dry cleaning businesses, CRAs might be developed for a number of options: (1) regulatory imposition of stringent housekeeping requirements; (2) development of new process technologies that reduce emissions; (3) development of new process technologies that use different, presumably more benign, solvent systems; (4) determination of practices that reduce the use of solvent systems from the beginning (e.g., data indicate that about half of all clothes sent to dry cleaning establishments are simply sent for pressing, so this service can be broken out as a separate offering); and (5) creating clothes and textiles that don't require dry cleaning at all.

3. Each CRA should be *monitarized*. While this step is difficult and raises significant concerns for many people, the alternative is to continue the current situation where non-monitarized variables simply don't get considered. On balance, therefore, this is probably a necessary step, although it requires that the assumptions and valuations inherent in the CRA analysis be very transparent and represent consensus to the fullest possible extent.

4. Once monitarized, the CRAs should be *integrated into the cost/benefit analyses*. In some cases, this will require that cost/benefit analyses also be expanded to include consideration of all options, which is a salutary practice in any event.

This task is not conceptually unachievable. It may be accomplished, for example, by defining a process similar to the CRA that can be developed for economic impacts —and for that matter cultural and other impacts as well, assuming one is able and willing to express all the results in comparable units (dollars, for example). The quantification of the impacts inherent in the equations can be derived from willingness to pay surveys or similar economic methodologies.

For economic impact, one can write:

$$E = eNP(d_i)$$

where E is economic impact, and e is a weighting factor reflecting the fact that the dollar value of economic impacts may be subjectively assessed at more or less than 1 by the public when compared with other values. Also note that economic impacts occur over time; this formulation assumes present discounted value is used, or, in other words, that the integration over time has been performed in the underlying assessment.

Similarly, for cultural values, the expression:

$$C = cNP(d_i)$$

can be defined.

At this point, a Comprehensive Policy Support Assessment (CPSA), which integrates the CRA, economic, and cultural considerations, can be defined for each option under consideration. For the jth policy option, for example, one would obtain:

$$CPSA_j = B_j + A_j + G_j + E_j + C_j$$

or, equivalently:

$$CPSA_j = CRA_j + E_j + C_j$$

The above methodology implies a quantitative process. In many cases, this may be unrealistic: Resources may not permit such detailed analysis; options may be difficult to define with sufficient specificity to support a quantitative assessment; data may be so sparse and uncertain as to render quantification misleading in itself. Moreover, the aggregation of very different kinds of data, much of it value laden, will inherently be difficult and controversial.

Even if only qualitative information is available, however, there is still value in following such a methodology, for it at least ensures that the implications of the decision under consideration for the system within which it will be implemented are being considered at some point in the policy development process. Moreover, the need to develop CRAs, or a family of CRAs, and integrate them into cost/benefit analyses using a comprehensive methodology such as the CPSA requires the explicit affirmation of the underlying linkage between environmental science, and economics and policy-making. This is a critical achievement in itself.

REFERENCES

Allenby, B.R. "Industrial ecology and comprehensive risk assessment." Chapter 17 in R. Kolluru, S. Bartell, R. Pitblado, and S. Stricoff, eds. *Risk Assessment and Management Handbook.* New York, McGraw-Hill Inc.: 1996.

Cohrssen, J.J. and V. T. Covello. *Risk Analysis: A Guide to Principles and Methods for Analyzing Health and Environmental Risks.* Washington, DC, U. S. Council on Environmental Quality: 1989.

Cothern, C.R., M.A. Mehlman, and W.L. Marcus, eds. *Risk Assessment and Risk Management of Industrial and Environmental Chemicals.* Princeton, Princeton Scientific Publishing Company: 1988.

Daly, H.E. and J.B. Cobb, Jr. *For the Common Good.* Boston, Beacon Press: 1989.

EPA (U.S. Environmental Protection Agency). *Unfinished Business: A Comparative Assessment of Environmental Problems.* (1987).

Gold, L.S., T.H. Slone, B.R. Stern, N.B. Manley, and B.N. Ames. "Rodent carcinogens: setting priorities." *Science* 258: 261–265 (1992).

Graedel, T.E. and B.R. Allenby. *Industrial Ecology,* Upper Saddle River, NJ, Prentice Hall: 1995.

Ryding, S., B. Steen, A. Winblad, and R. Karlsson. *The EPS System: A Lifecycle Assessment Concept for Cleaner Technology and Product Development Strategies, and Design for the Environment.* Avfallsforskningsradet AFR, 1993.

Steen, B. *EPS—Default Valuation of Environmental Impacts from Emission and Use of Resources (Version 1996).* Avfallsforskningsradet AFR, 1996.

Zeckhauser, R.J. and W.K. Viscusi. "Risk within reason." *Science* 248: 559–564 (1990).

EXERCISES

1. You are an environmental regulator in a developed country faced with a decision as to whether to permit mining in a wilderness area that probably contains threatened species. You are required to perform an assessment of the desirability of this activity.

   a. What should your option set be?

   b. Perform a CPSA for each option (this can be qualitative). What stakeholders should you involve in order to support your CPSAs? What are the major issues that can be resolved by gathering data, and what issues involve value judgments?

   c. Are there any issues you feel are important, but you cannot fit into a CPSA? Assuming that there are, how would you ensure they are considered as part of the regulatory process?

2. You are the Minister of Science, Technology, and Environment in a rapidly industrializing Asian nation reviewing a cost/benefit study prepared for an industry group on some proposed clean air regulations. The study indicates that implementing such regulations would impose significant costs on the economy. Using the questions set forth in this chapter, what concerns might you have about the conclusions, and what assumptions might you pay particularly close attention to?

3. You are in charge of the Energy Directorate in a rapidly developing Asian nation. If you do not implement rapid growth in energy production, your analysts tell you that you will reduce your nation's growth rate by approximately 2 percent per year, which could destabilize the government. Your analysts also tell you that nuclear power appears to be the only option that can provide the amount of energy required within the appropriate time frame. On the other hand, Aswame, a neighboring nation with which you have fought two wars in the past decade, has threatened to develop a nuclear weapons capability if you develop nuclear power plants because it fears that your nation will divert nuclear fuel to weapons use. Moreover, you have a small but growing vocal minority that is against anything nuclear.

   a. Given conditions in your country, develop a set of energy production and consumption options that might be available to you.

   b. Use the CPSA methodology to organize your thoughts on the costs, benefits, and risks of each option.

C H A P T E R 1 1

Economic Issues

11.1 OVERVIEW

There are many critical dimensions to the policy issues arising from the integration of environment, technology, and economic activity throughout the global economy, such as trade and national security (addressed in Chapter 18). Underlying these dimensions, however, are two disciplines that tend to predominate in policy formulation and implementation. These are, of course, economics, the subject of this chapter, and law, covered in the next chapter. The importance of these disciplines for progress toward sustainability cannot be over-emphasized; indeed, it is safe to say that a desirable sustainable world cannot be achieved until and unless society gets the economics right. Whether this can be accomplished, or how, remains controversial at this point.

Economics is perhaps the most powerful discipline in terms of its capability to shape policy. Economic analysis tends to drive, or at least strongly influence, most national and international policy formulation. Policies and institutions intended to foster economic development, for example, are guided by economic analysis and carried out by economic means by institutions such as the World Bank and the International Monetary Fund. Indeed, the status and success of nations is frequently judged by economic indicators such as per capita income or Gross National Product (GNP). Although it is a unique example, Japan has grown to become a global power based on its economic performance, in spite of a Constitution that prevents it from projecting its military powers beyond its own borders.

Economic systems exist within, and give structure to, different social and cultural systems, and many of their characteristics can only be understood within their cultural context. Thus, for example, even though most developed countries have market economies characterized by private firms, the structure and implicit role of such firms differ greatly among the United States, Japan, and France.

There is a large and growing literature dedicated to the difficult issues arising from the effort of many disciplines, particularly economics, to adjust to the increasing importance of environmental considerations. Especially in the case of economics, this has involved challenges to fundamental assumptions and methodologies. This is not surprising. Most of the intellectual capital of the modern world has been accumulated during a period, the Industrial Revolution, when technological and organizational evolution created essentially unlimited material and energy resources, as demonstrated by the explosive growth of human population during the period (recall Figure 3.5). Accordingly, there has been little pressure to consider global systematic constraints on resources and sinks, as opposed to local allocation questions, in developing such fields as economics (and law, for that matter). From this perspective, then, it is natural that assumptions appropriate to a period of relatively unlimited growth, not evolution toward sustainability, are implicit in such disciplines. Thus, for example, the neoclassical position that continued economic growth (as opposed to development) is necessary for full employment assumes no environmental limits to such a process, and an economic/legal structure that provides no representation in resource allocation conflicts for future generations assumes that such representation is unnecessary because they will have what they need when they need it regardless of current resource consumption decisions. Both positions are valid if continuing unlimited global growth in population, economic activity, and resource consumption can be assumed. Both positions become problematic if this assumption is inappropriate.

Obviously, this text can provide only an overview of some of the principle economic issues arising in this area; there is a large and growing literature to which readers should turn for more detailed discussion. Nonetheless, such issues raise many important policy considerations, and an introduction to them provides a critical understanding for the student of industrial ecology. Perhaps equally important, they illustrate the point that the intellectual tools used to develop and deploy policy may themselves be part of the problem, and thus they, as well as the problem under consideration, need to be critically evaluated.

11.2   ELEMENTS OF ECONOMIC THEORY AND PRACTICE

The interactions between economics and industrial ecology are in the initial stages of being explored; it is likely that both will change to reflect insights gained from the other over time. The study of industrial ecology, which will support the evolution of environmentally and economically efficient economic and technological systems, obviously must draw on economic expertise. Similarly, economic theory that is divorced from environmental externalities and constraints will become increasingly inappropriate for informing policy development. Accordingly, this summary should be regarded as preliminary and somewhat impressionistic; the growing literature in the nascent sub-discipline of ecological economics provides a more detailed treatment of many of these issues. Even at this early point, however, it is possible to identify certain problematic issues of which the student should be aware.

11.2.1   Issues of Scale

Standard neoclassical economics holds that economic growth is necessary to maintain full employment. If true globally, this obviously raises a number of difficult issues. Initially, it is important to distinguish between economic *development*, where quality of life is maintained or increased even though the physical basis of the economy—consumption of energy, materials, and the absorptive capacity of sinks—remains the same, from economic *growth*, which implies continuing increases in resource consumption.

The former is to be desired; indeed, it is inevitable: Economies, like all complex systems, change and evolve; that is their nature. That economic development can continue without growth in material consumption is suggested, but certainly not proven, by the growing substitution of information for other inputs into the economy (refer to Section 6.7.1).

Another interesting potential interaction involving economic theory, the information revolution, and sustainability issues has been raised by Graciela Chichilnisky of Columbia University, who suggests that traditional economic analysis, which is concerned primarily with the creation of wealth and not the fairness of its distribution (or "distributive equity"), breaks down when knowledge becomes an important economic input (footnotes omitted):

> Markets with knowledge [as an important input] are markets with privately produced public goods, and these behave quite differently from classical markets. A newly discovered fact is that in these markets efficiency and distribution are very closely interlinked, in a distinctive way. For efficiency, the distribution of property rights must be relatively more egalitarian, assigning more public goods to those who own fewer private goods. This is in stark contrast with conventional markets where efficiency and distribution are divorced from each other.
>
> In the new society based on knowledge, markets may require a more equal distributions [sic] of wealth to function efficiently. The knowledge society could be more egalitarian than the industrial and the agricultural societies, although this is only a well-informed hope requiring further investigation.

The possibility that sustainable development, with its strong equity dimension, can be linked to greater economic efficiency as a result of the information revolution is intriguing, and could be an important basis for developing policies that align self-interest with both economic growth and economic efficiency.

Whether the natural systems within which all economic activity takes place can support continuing physical growth of the economy is strongly questioned by many environmental scientists. The adequacy of the resource base, including atmospheric, oceanic, and soil sinks, to support a world of 8 to 14 billion people, all living as well-off people in developed countries do now using current technologies, is problematic. Whether human ingenuity can shape new technological and cultural systems that can accommodate development without such growth—in part by delinking resource consumption from quality of life—is a different question.

It is not that standard economic analyses, with their assumptions of virtually unlimited resources, are necessarily wrong when properly applied and scaled (in many microeconomic assessments, for example). Rather, it is that care must be taken to ensure that such analyses are used where such assumptions are appropriate. This is particularly true when policies have potentially substantial impacts on regional or global natural systems or resource consumption patterns, and thus the unlimited resource assumption may be invalid.

11.2.2 Issues of Scope

Economics is generally viewed, by most economists at least, as an objective, not normative, discipline. This view has been strongly challenged by others. As Robert H. Nelson notes, "Economics is not only a technical subject; it also reflects a strong set of values ... often at odds with the way of thinking of biologists, ecologists, and other physical scientists." Thus, for example, physical systems such as wetlands are considered "worthwhile" to the extent they are economically efficient, a term that in many cases is sufficient in itself to provide presumptive social legitimacy.

In particular, it is an important assumption in neoclassical economics that everything has a monetary value; that is, that there is literally nothing sacred. While this assumption may be necessary to develop internally coherent quantitative methodologies, it is obviously contentious, and at the level of the discipline itself, a powerfully normative statement. It leads, for example, to assertions that, if people are not willing to pay for the preservation of a species, the species is worth nothing and its disappearance is immaterial. Such statements tend to generate significant friction between some economists and environmentalists, as the latter tend to have strongly held values concerning environmental issues. As in the case of risk assessment, the inclination of experts (in this case, economists) to achieve "objective" results runs afoul of the public's strong belief that subjective dimensions are meaningful in such situations.

A related issue arises because modern economics tends to be highly quantitative. Thus, a number of valuation methods have been developed to quantify such difficult phenomenon as the health effects of pollution. Examples include the following:

- The *human capital* method, which measures earnings foregone due to illness or premature death as a result of pollution exposure.
- The *cost of illness* method, which measures lost workdays plus out-of-pocket medical and associated costs resulting from pollution exposure.
- The *preventive/mitigative expenditure* method, which measures expenditures on activities to mitigate or reduce the effects of pollution, such as putting in new water delivery systems to avoid exposure to contaminated groundwater.
- The *wage differential* method, which uses wage differentials between areas differing in pollution exposure as a surrogate for the implicit value of less pollution for people.
- The *contingent valuation* method, which uses surveys to determine what value people say they put on pollution avoidance.

Obviously, all of these methods have some drawbacks and confounding factors, such as the well-known difficulty of associating verbal responses to questions with subsequent behavior. Nonetheless, they offer a means by which dollar values can be assigned to environmental insults, which can then be factored into quantitative analyses.

Although such an approach simplifies analysis and can generate more rigorous and understandable results, it can also mean that factors that cannot be quantified are, in practice, simply not included in the analysis. Even when qualitative impacts are considered, they tend to be assigned lesser weight than the former. Consider, for example, a project that has certain economic advantages, but that will result in ecosystem disruption and perhaps species extinctions. Most economic analyses will not include the latter; even if it is included, it will be quantified on a "willingness to pay" or potential market value basis (as a source of drugs, for example). It will not include the nonquantifiable element: The value of the species for its own sake.

A particular difficulty with quantification arises because of the nature of many environmental perturbations. Quantification of the impacts of these perturbations is frequently impossible, both because of lack of data and powerful enough models and because of the inherent nature of the system itself (many such systems are deterministically chaotic, which means that even in theory their evolution over time as a result of human forcing cannot be accurately predicted). It may well be impossible to know, for example, what the costs of global climate change might be until it actually occurs, especially at local and regional levels. The system is simply so complex, nonlinear, and potentially discontinuous that accurate quantification of future costs attributable to current activities is essentially impossible. This fact has led some economic analysts to ignore potential discontinuities, and to treat the predictable costs of global climate change as minor, a conclusion regarded by many climate scientists as unrealistic. It is seen as deriving from the limitations of economic analysis, rather than a reasoned understanding of the actual dynamics of the system.

As with scale issues, these concerns, even if valid, do not invalidate modern economics. Such analyses have proved their worth in informing policy many times over. What they do indicate, though, is that the appropriate scope of economic analysis may be bounded, and that such techniques should be used with caution where the moral or ethical content of the issue is significant, and where uncertainties and the nature of the systems under consideration have important qualitative dimensions.

11.2.3 Discount Rates

Standard economic analysis (in part probably reflecting human psychology) asserts that money today is worth more than the same amount of money tomorrow, based on inflation and the returns over time that can be anticipated if the money is invested. To reflect this, a "discount rate" is applied to future returns as compared to current returns. Technically, this is represented by an equation that gives the present value, A, of an amount V, which will be available t years from now; it is apparent that A is smaller than V:

$$A = V \, (1 + i)^{-t}$$

where i is the discount rate.

The use of such discount rates to value resources and plan investments in business and government is ubiquitous and, in many cases, appropriate. Without such an approach, it would be difficult to compare investments that required expenditures and generated streams of returns in differing time periods. Obviously, however, this approach also provides a strong incentive to use resources as soon as possible, rather than save them for the future. If the return is properly invested so that it provides a stream of benefits over time, the future may indeed benefit more from the economic growth generated than it would have if the resources were conserved (assuming no unique features about the resources themselves, which tends to imply renewable in the short term). Usually, however, there is no consideration of the implications of discounting in terms of eventual resource limitations or other externalities that may be associated with such growth. For example, even though trees are generally renewable, logging an old growth forest destroys a unique ecosystem, which cannot be reconstructed.

Moreover, the assumption that returns are properly invested may be too strong, especially in developing countries that are selling resources. In many cases, the elites in such countries simply appropriate any returns from resource depletion, and consume rather than invest them. Indeed, it is worth noting that there are some indications that even the non-environmentalist public does not support the argument that, rather than spend money now on environmental improvements, society should invest in the short term, and fix environmental perturbations in the long term with the resulting returns. Thus, Kempton et al. in their in-depth survey of American environmental attitudes got the following results (figures are positive responses to question in percent):

Question 28 We should invest in industry rather than spending money on the environment, so that our economy will grow. Our children would then be more prosperous and better able to afford the cost of fixing any environmental problems we may have caused.

Earth First!	Sierra Club	Public	Dry Cleaners	Sawmill Workers
0	0	25	14	15

In general, the discounting concept, if applied universally in an economy with significant externalities as is the case today, is incompatible with the fundamental industrial ecology principle that the economy should function so as to be sustainable over a relatively long time period. It leads to the elevation of short-term parochial factors over longer term resource management concerns. In a famous case in California, for example, a timber company was taken over in a process that generated a need for rapid payoff of so-called "junk bonds," bonds paying a very high interest rate to the holders. Accordingly, the firm, which controlled substantial stands of increasingly unique virgin redwoods, immediately began to cut them as rapidly as possible to pay off the loans. In effect, the junk bonds generated a very high discount rate, which made the trees much more valuable as they were more rapidly exploited.

Nonetheless, it remains true that in many cases discount rates are a useful planning tool, so the question is not one of rejecting the concept outright, but of understanding under what conditions such analyses are useful. In particular, discount rates should be applied cautiously in cases where significant social value issues and externalities are present (as in the California case), or where nonrenewable resource consumption is involved, especially in cases where it is apparent that investment of resultant revenue streams is unlikely.

11.2.4 Substitutability versus Complementarity of Resources

Another assumption common to standard economic analysis is substitutability among resources and economic inputs based on monetary value, which in turn reflects relative scarcity. In general, this is a valid principle that has been demonstrated many times: as one input becomes scarce, and thus more expensive, another is substituted for it. The difficulty with this principle, then, arises not because it is wrong, but because it is right so much of the time that it has become axiomatic, rather than an assumption. As applied in practice, in other words, it assumes that there are always substitutes. It has become an embodiment of unquestioning technological optimism. It is in this latter guise that it is questionable.

For example, if the electronics industry were to substitute indium solders for lead solders in electronics products, it would rapidly draw down world reserves of indium. Given current practices, as the electronics products are discarded, the indium would arguably be rendered unrecoverable. Should indium then turn out to be a critical, and unique, material for some socially important product—room temperature superconductors, for example—the future would indeed have been deprived. Indium, in other words, would be a complement to other materials in a critical technology, not a substitute. Another example is the element gadolinium, which has the property of heating up when placed in a magnetic field, and has thus been used to develop super efficient refrigerator technology (a conventional refrigerator has a maximum efficiency of about 40 percent; a gadolinium refrigerator has a theoretical efficiency of about 60 percent). Given the burgeoning demand for refrigeration, especially in rapidly developing tropical countries, and the lack of other materials with this property, gadolinium could well be a complement to efficient cooling, with no (as yet identified) substitutes.

The substitutability versus complementarity debate comes down in many cases to assumptions about technology and technological evolution. If one is a technological optimist, the tendency is to assume substitutability; if one is a technological pessimist or has ideological predispositions against technology, the tendency is to discount substitutability. While the history of the Industrial Revolution clearly supports the power of technology to evolve substitutes in most cases, it does not justify the strong axiomatic formulation.

This is particularly the case as environmental constraints and concomitant regulations, such as bans on certain materials, reduce the design space of future technologies. Under these circumstances, substitutability will become harder and harder to achieve. Thus, as above, it is not that the assumption is itself wrong; rather, caution must be exercised in its application.

11.2.5   Externalities

An externality is simply a cost, either positive or negative, that is not captured within the economic system through prices. Thus, for example, when a factory is allowed to dump toxic materials in a river for free, it generates a negative externality to the extent that it imposes costs on society that it doesn't have to pay. When fish die downstream, the local fishermen are indeed poorer than they otherwise might have been, but the price paid by the factory to dump its waste does not reflect this cost.

Externalities matter because much of the purpose of environmental management and regulation is to compensate for externalities, often in less efficient ways than simply changing prices. For example, post-consumer product takeback regulations may be seen as an effort to internalize to the manufacturing firm the environmental costs associated with management of the environmental impact of used products. Clean air laws that prescribe certain control technologies may be seen as efforts to internalize to the firm the costs associated with the emissions being controlled.

As has been pointed out previously in this text, properly bounded free market systems are preferable to any form of centrally controlled system in informing and achieving economically and environmentally efficient performance. It is also analytically defensible, as Nordhaus pointed out in 1992 in an article on the ecology of markets, to argue that internalizing externalities in a market structure is, in itself, efficient in controlling environmental impacts. As he notes, however, there are a number of barriers that make this theoretical achievement virtually impossible:

1. Political opposition to externality taxes or fees (the easiest way to internalize externalities, assuming valuation problems can be resolved).
2. Difficulty in determining quantitative values for externalities, which by definition have no market within which they are priced (and, as noted above, often involve issues of morality and the sacred, which many people find difficult to quantify under any circumstances).
3. The extreme heterogeneity of some externalities, which vary continuously over different spatial and temporal scales, making any quantification both difficult and potentially arbitrary as cost data are aggregated.
4. "Natural monopoly" distortions of the market, where the externality is a public good, and cannot be "privatized" into the private market (environmental security aspects of national defense might be such an example).

In addition to these barriers pointed out by Nordhaus, there is a fundamental systemic barrier to the simple path of internalizing externalities. In many cases, the natural systems that are involved—atmospheric, oceanic, and biological systems, among others—are deterministically chaotic, which means that their evolution through time cannot be accurately predicted. This means, of course, that the costs resulting from that evolution also cannot be predicted, but can be quantified only as they actually occur. By that time, it is obviously impossible to change prices to prevent the costs from accruing: They have already done so.

11.2.6 Rational Agents

In order to model consumer behavior (and many other economic factors as well), modern economic analysis relies on the concept of utility. *Utility* is, in brief, the ability of a good or set of goods (or investment(s), or activity(ies), or whatever) to satisfy wants. In order to model utility, and thus demand, rational consumer behavior must be assumed; as Watson and Holman point out in their text on price theory: "To say that the consumer behaves rationally means that the consumer calculates deliberately, chooses consistently, and maximizes utility. Consistent choice rules out vacillating and erratic behavior." Predictable demand is a prerequisite to much of the mathematics that underlies economic theory. Needless to say, for years this assumption of rationality has been challenged as being grossly out of kilter with observable human behavior. Modern theory, therefore, incorporates concepts such as "expected utility of behavior under uncertainty" and the "rational expectations" hypothesis, rather than strict rationality assumptions.

Nonetheless, even these less stringent requirements have been challenged based on evidence that people, in fact, will choose differently among the same options if the options are presented in a different manner, and that non-rational psychological and cultural models dominate much decision making. Moreover, there is the obvious difficulty that agents have very different reasoning capabilities (they are heterogeneous), and for many purposes they rely not on rational assessment, but on the application of heuristics that make life simpler (a damaging heuristic, for example, is racism: it reduces the need for thought, albeit at a high personal and social cost).

The role of cultural models and values is particularly important in considering environmental issues, where, as Kempton et al. found, they form a powerful filter for both obtaining information and developing personal positions. In interviews with experts on global warming, for example, they found (122–123) that each specialist—legislative aide, automobile engineer, environmental advocate—was, in fact, "living in different worlds of information sources":

> The assimilation and use of information by these participants [in the in-depth interview process] is influenced by their own mental models, values, and political ideology . . . . We find several instances in which these individuals actively search for information that fits the constraints imposed by their ideology and their employer . . . . As a result, these individuals differ on basic facts. (163) We found multiple processes that select and filter information so as to buttress ones own (or one's organizations') ideology and interests. (186)

The argument that internalizing externalities—getting prices right—will solve environmental issues through operation of market mechanisms is probably true in general direction. But it assumes rational agents in both valuation and efficient market functions. It assumes, in short, a *deus ex machina* by the Invisible Hand, optimizing not just current welfare, but welfare into the future. This is particularly difficult, where, as Wilde et al. point out, "[t]he ability of persons and societies to act 'rationally' in both the economic and neurophysiological senses is decreasing due to the growth of knowledge." Unfortunately, it seems more likely that the assumption of (even limited and bounded) rationality of economic agents is difficult to apply under these circumstances, and, therefore, the obligation of society to make difficult choices remains. To quote Wilde et al. once again:

> . . . institutions must evolve which enable each agent in the society to know less and less about the behavior of other agents, and about the complex interdependencies generated by their interactions . . . . More than ever, and with increasing speed and intensity, cultural and institutional evolution must be conscious and deliberate. Man lives more and more in a world of his own making and can no longer rely on everything working out for the best.

Although not intended to apply to industrial ecology and the complex systems with which the field deals, its sentiment could not be more apropos.

11.2.7   Static versus Dynamic Analysis

Many economic models, particularly older neoclassical ones, are static and assume the existence of an equilibrium to which the system, once perturbed, always returns; as Richard Nelson says in his book *The Sources of Economic Growth* (4), "[c]onventional economic theory stresses the primacy of for-profit firms, in competition with one another, operating in markets in which supply and demand are balanced so as to determine equilibrium prices and quantities." Quoting Shumpeter—"The essential point to grasp is that in dealing with capitalism we are dealing with an evolutionary process"—Nelson, like an increasing number of his compatriots, questions this approach (8, 3–4, 15):

> ... the standard approach is problematic in] at least three respects .... One was its proclivity to "divide up the sources of growth," when in fact there was powerful evidence that they were strong complements. A second was a tendency to treat economic growth as a process involving moving, but continuing, equilibrium, whereas evidence of continuing disequilibrium was very powerful. Third, much of the institutional complexity of modern capitalism was repressed ....
>
> ... in a regime of continuing technical advance, the economy as a whole is in a continuing state of disequilibrium. [I present] a different picture [than the static neoclassical one], one in which technological advance and economic growth are seen as proceeding through the operation of a complex set of institutions: some for-profit, some private but not-for-profit, and some governmental .... Institutional change, like technological change, must be understood as an evolutionary process. The result is that modern capitalism is a very complex system.
>
> More generally, the assumptions built into the simple form of the neoclassical model —that technological knowledge is a public good and that growth is an equilibrium process —would appear to be inconsistent with the mechanisms that draw forth new technologies in capitalist economies.

Given the importance of technological evolution for achieving greater economic and environmental efficiency, and the dominance of economic analysis in policy formulation, the possibility that there are fundamental flaws in the mental and mathematical models applied by mainstream economists to these issues should be a matter of serious concern to policymakers.

At least in one subtle way, the dominant economic mental model is probably hindering effective policy development. By encouraging an atomistic view of the firm, it hinders the development of policies intended to take advantage of their internal heterogeneity, and unintentionally supports an ideology of firms as evil entities, rather than complex systems in their own right.

More recently, questions arising neoclassical assumptions have been raised by a few researchers developing and applying "agent-based" computer modeling techniques to the study of social phenomenon such as economic behavior and trade in "artificial societies." The results of this nascent experimental work, as described by Epstein and Axtell in their recent book *Growing Artificial Societies: Social Science from the Bottom Up* (137, footnotes omitted), echo the words of Nelson above:

> The emphasis in the economics literature has been on the *existence* of static equilibrium, without any explicit microdynamics. Why cannot prices oscillate periodically on seasonal or diurnal time scales, or quasi-periodically when subject to shocks, or even chaotically? Is it not reasonable to expect generational or other long-term structural shifts in the economy to produce prices that follow a trend as opposed to staying constant? Might not far from equilibrium behavior be a more reasonable description of a real economy? From the computational evidence [produced by the authors], we think that there is good reason to be skeptical of the predominant focus on fixed-point equilibria. Economies of autonomous adaptive agents—and of humans—may be far from equilibrium systems. And, in turn, far from equilibrium economics might well turn out to be far richer than equilibrium economics.

As Epstein is quoted in a *Science News* article, "The assumption that we can let markets produce efficiency allocations [of capital or resources] on their own is deeply challenged by our work . . . . We see how brittle traditional economic theory really is."

It is perhaps too early to tell whether criticism by historical economists such as R. R. Nelson, and the results of artificial society modeling, augment or invalidate standard neoclassic static analytical methods or results, and, if so, under what conditions. Nonetheless, the implications of the work—that economic systems are complex, dynamic systems, and technological change is best understood in that context—are important for rational industrial ecology policy development, and reinforce the importance of understanding the nature and dynamics of the systems with which one is concerned.

11.3 LABOR IMPACTS

Any period of economic transition, even if it increases productivity, output, and GDP —or, for that matter, environmental efficiency—will have disruptive effects on labor markets, at least in the short term. Even flexible labor markets, such as in the United States, take time to change. New skills are not developed overnight, and neither firms nor individual employees adapt rapidly to change. Thus, for example, the post-Civil War industrialization of the United States displaced many craftsmen, small manufacturers, and small retailers even as it led to unprecedented economic expansion. Even earlier, the advent of the Industrial Revolution in England gave rise to the violent reaction of the Luddites, who smashed power looms and spinning jennies, technologies that were perceived as causing widespread unemployment. It is thus reasonable to ask what impacts on employment moving the economy towards sustainability might have, and how such impacts might be reduced.

The little work that has been done in this area is not persuasive, in part because of severe definitional questions and the inevitable uncertainty associated with projections of technology and employment in a period of rapid change. Several relevant observations, however, may indicate that, over the long term, there need not be any significant impacts on employment resulting from the achievement of a more economically and environmentally efficient economy, particularly if informed policies are initiated. In this regard, recall that it is likely that a more sustainable economy will see an increase in services and a decrease in manufacture as industry increasingly sells function, rather than specific products, to customers. What is the record of such service-based economies today?

First, it is generally not the case that technological change reduces employment or economic growth; historical data indicate the opposite, in fact. Moreover, the data do not support the more specific claim that a service economy produces less jobs than a manufacturing economy. Indicative of this is the experience of the United States, which has an economy heavily oriented toward services, yet historically maintains a low unemployment rate compared to other developed nations, particularly in Europe. Additionally, it appears that there is little ground for the claim that service jobs are low-paying, low-quality jobs ("McJobs"). Rather, many of the jobs added in such economies are professional or managerial. To the extent current experience with service sector oriented economies is relevant, therefore, the long-term employment outlook in a sustainable economy may be relatively good.

In the short term, however, it is likely that at least transitional unemployment levels may rise, particularly in countries with relatively rigid labor markets (overly burdensome restrictions against firing workers and closing facilities, for example) and product-market barriers (e.g., restrictive zoning laws or legal limits on shop-opening hours). It is also likely that a greater premium will be placed on skilled labor, with the unskilled laborer, particularly in developed countries, in a difficult position indeed. (The definition of "skilled" might well shift, however, from "skilled in manufacturing" to "skilled in knowledge work".) This, of course, implies that government policies that reduce rigidity in labor and product markets, and support the development of a skilled labor force, will reduce unemployment even in a period of rapid evolution toward a more sustainable economy (these tend to be desirable policy goals for a number of reasons). Less obviously, a research program that helped the government, private firms, and academia understand the evolution and structure of a more sustainable economy, and thus what skills such an economy might need, could help to reduce such impacts perhaps significantly, although it could not be expected to obviate them completely.

11.4   FINANCE, CAPITAL, AND INVESTMENT

In traditional Keynesian economics, investment is simply aggregate income less consumption, and, at equilibrium, is equal to savings. The importance of investment for industrial ecology policy from a traditional perspective is apparent: If progress toward a more sustainable global economy requires rapid technological evolution, then it must be financed, and technological change will be financed if it can attract investment. It will attract investment if financiers and industrial organizations believe that, over time, they will earn an adequate return on the investment.

Investment and technological evolution are not just synergistic, but inextricably intertwined. Thus, for example, one reason given for the success of Silicon Valley, perhaps the most innovative region in the world, is the unparalleled access to venture capital in that area, even compared to other American regions such as the "Route 128" corridor around Boston, Massachusetts. Conversely, in many European financial markets, as well as in Japan, capital is much harder for start-ups to obtain, so there are correspondingly fewer small firms with their concomitant innovation. The figures required for the diffusion of new technologies can be daunting. For example, one estimate by the World Business Council is that bringing the paper products industry up to best industry standards would take a worldwide capital infusion of at least 20 billion dollars.

Establishing a robust financial market, and being able to attract adequate capital, is a complex problem, especially for developing countries, which must in most cases rely on foreign investment. In earlier years, much of this foreign investment was from developed country governments in the form of foreign aid, or from multilateral lending organizations such as the World Bank. More recently, however, the bulk of such investment has come from the big transnational firms in the form of FDI (foreign direct investment). The amounts involved can be significant. In 1994, some 90 billion dollars (U.S.) flowed into developing countries; FDI in China alone for the past several years has amounted to some 30 to 40 billion dollars (U.S.) annually (Figure 11.1). FDI is important because it not only helps create the capital base for indigenous industry,

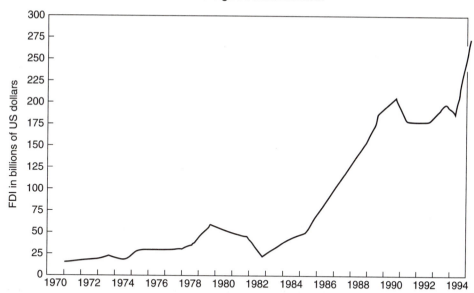

FIGURE 11.1   Foreign direct investment from major source countries (Belgium, Canada, France, Germany, Italy, Japan, the Netherlands, Sweden, Switzerland, the United Kingdom and the United States), 1970–1994.
Source: Based on International Monetary Fund, *International Economic Outlook*: May 1997.

but because it supports the transfer of modern technology—which is usually more environmentally as well as economically efficient than the alternatives—as well as modern business practices and management. Thus, it is perhaps no surprise to see that FDI corresponds closely to GDP for the developing world (Figure 11.2).

This matters for several reasons. First, it implies that one benefit of a globalizing economy is that technological improvements, once adopted by world class transnationals, may be anticipated to diffuse to developing as well as developed countries as part of normal global investment patterns. Second, it focuses again on the critical role of industrial firms as agents not only of development but, at least potentially, of environmental efficiency and even sustainability. (Some of the problems this nascent role may raise are discussed in more depth in Chapter 16.) Third, it affirms a strong potential link between financial institutions and practices and environmental performance: FDI and associated multilateral lending streams are important mechanisms by which environmentally preferable practices can be diffused. Finally, a basic point is that technological evolution, whether it is in developed or developing countries, must be financed, and it will only be financed if it is profitable in a competitive capital market.

One additional financial activity is worth noting: The growth of so-called microfinance institutions, which provide small loans ("microcredit") to borrowers without collateral, usually women, who are members of a "peer" or "solidarity" group. The group guarantees the payback of the loans for its members. Default rates on such loans are small, usually well under 3 percent. While the concept is not new—analogous small credit unions existed in Germany in the last century—the extent to which it is being implemented in developing countries is. Some of the Asian institutions active in this

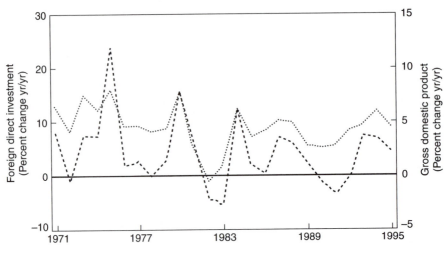

FIGURE 11.2 Investment and GDP in the developing world, 1971 through 1995. The two tend to track each other with little apparent lag time.
Source: Based on the Conference Board, *StraightTalk*, May 1997.

area have over two million borrowers each. The effects of such activity not only reduce poverty, but empower women, which in turn affects population pressures—yet another example of the linkages of capital markets and sustainability.

REFERENCES

Anderson, F.R., A.V. Kneese, P.D. Reed, R.B. Stevenson, and S. Taylor. *Environmental Improvement Through Economic Incentives.* Baltimore, Johns Hopkins University Press: 1977.

Bruchey, Stuart W., ed. *Small Business in American Life.* New York, Columbia University Press: 1980.

Cairncross, F. *Costing the Earth.* Boston, Harvard Business School Press: 1992.

Chichilnisky, C. "The knowledge revolution." *Columbia University Discussion Paper Series No. 9697-06,* November 1996.

Costanza, R., ed. *Ecological Economics: The Science and Management of Sustainability.* New York, Columbia University Press: 1991.

Daly, H.E. *Steady-State Economics.* Washington, DC, Island Press: 1991.

Daly, H.E. and J.B. Cobb, Jr. *For the Common Good.* Boston, Beacon Press: 1989.

The Economist. A survey of the world economy (center section, pp. 1–46), September 28, 1996.

Epstein, J.M. and R. Axtell. *Growing Artificial Societies: Social Science from the Bottom Up.* Washington, DC, Brookings Institution Press: 1996.

Hirshleifer, J. "The expanding domain of economics." *The American Economic Review* December 1985, pp. 53–68.

Kempton, W., J.S. Boster, and J.A. Hartley. *Environmental Values in American Culture.* Cambridge, Ma., The MIT Press: 1996.

Kneese, A.V. *Measuring the Benefits of Clean Air and Water.* Washington, DC, Resources for the Future: 1984.

Machina, M. J. "Choice under uncertainty: problems solved and unsolved." *The Journal of Economic Perspectives* 1(1): 121–154, Summer 1987.

Nelson, R.H. "Sustainability, efficiency and god: economic values and the sustainability debate." *Annu. Rev. Ecol. Syst.*, 26:135–154, 1995.

Nelson, R.R. *The Sources of Economic Growth.* Cambridge, MA, Harvard University Press: 1996.

Nordhaus, W.D. "The ecology of markets." *Proceedings of the National Academy of Sciences* 89:843-850, February 1992.

Peterson, I. "The gods of sugarscape: digital sex, migration, trade and war on the social science frontier." *Science News* 50:332–333, November 23, 1996.

Stix, G. "Small (lending) is beautiful." *Scientific American* April 1997, pp. 16–20.

Wilde, K.D., A.D. LeBaron, and L.D. Israelsen. "Knowledge, uncertainty, and behavior." *AEA Papers and Proceedings* 75(2):403–408, May 1985.

EXERCISES

1. You are the Minister of Industry for a small developing country with extensive bauxite reserves, but few other resources and little manufacturing or service industry. A mining company has approached you and requested permission to mine the bauxite, which will be processed into raw aluminum on site.

    a. What issues does this request raise?

    b. What kinds of information would you need before deciding whether to approve the project?

2. Identify five cases of completely substitutable commodities or resources, and five cases of complementary commodities or resources.

    a. Briefly explain each selection.

    b. How, if at all, does your explanation depend on assumptions about time frames (for example, does a complement in the short term become more of a substitute in the longer term), or economic relationships (for example, relative prices)?

    c. Repeat this process for five cases where purchase of a function is substituted for purchase of a product. Are these cases more similar to substitution or complementation, or are these relationships an insufficiently sophisticated way to think about such cases?

3. Assume that global climate results in significant impacts on agriculture in California, and, further, that these impacts cannot be predicted with precision before they occur (because the system is technically chaotic).

    a. What policies would you as the California Governor recommend as economically efficient under the circumstances?

    b. How does your answer change if you are the President of the United States? If you are the Prime Minister of India?

CHAPTER 12

Legal Issues

There are obviously a number of legal issues raised by the transition of environmental issues from overhead to strategic for society. Some of these arise from the nature of law itself: Legal structures embody the culture within which they arise, and are therefore slow to change and are generally conservative. Thus, different policy concerns such as trade, limitation of monopoly power (antitrust), or consumer protection against fraud, for example, generate over time complex legal edifices to address them. When a society must deal seriously with a new set of issues that were not of concern as existing legal structures were established, such as environment in this case, it therefore is not surprising that conflicts arise as previously disparate policy structures must be integrated. Given the conservative nature of the law, resolving these conflicts is generally neither rapid nor uncontentious.

The way in which different jurisdictions implement their legal structures also reflects cultural values and morays. Some differences are historical: Countries that have been influenced by the British tend to have a "common law" system, while continental European countries, and others influenced by them, tend to have legal systems based on the Napoleonic Code. Even within these groupings, it is important to recognize and differentiate between informal and formal legal structures. The formal legal structures are those that are written down, while the informal ones are those characterizing the actual operation of the legal system in its society. In some countries such as the U.S., for example, written law is strongly respected, and the principle that "no person is above the law" is, for the most part, true. In other jurisdictions, the written laws are honored more in the breach, than in the observance, and connections of friendship and family are more meaningful than legal obligations. It is not unusual in some developing countries, for example, to find that industrial operations associated with the ruling elites are in practice held to less demanding environmental standards than similar facilities operated by transnational firms, regardless of written law.

It is also important to recognize some of the fundamental shifts occurring in legal systems around the world regarding environment. In this regard, two trends are important. The first is the explosive growth of environmental regulation in the past 20 years, as demonstrated not only by the growth of environmental statutes in developed countries such as the United States, shown in Figure 12.1, but in the growth of international agreements regarding the environment, which show an analogous growth (Figure 12.2). The second is the growth of environmentalist "non-governmental organizations,"

Growth in the Number of U.S. Environmental Laws

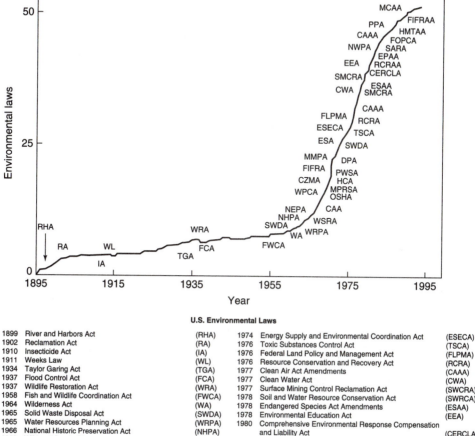

FIGURE 12.1   Growth in the number of U.S. Environmental Laws. Note the exponential growth of recent decades appears to be slowing, perhaps an indication of the maturing of first generation, command-and-control legislative tools.

or NGOs, which have become a de facto arm of international environmental compliance and monitoring activities, as well as increasingly powerful contributors to the development and enforcement of international environmental agreements. The 1995–1996 *Yearbook of International Organizations*, for example, lists some 36,054 non-governmental organizations, many of which are active on environmental or, more broadly,

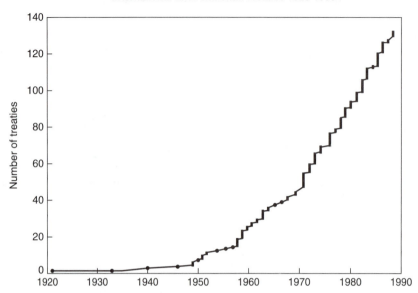

FIGURE 12.2 International Environmental Treaties, 1920–1990
Source: Based on N. Choucri, *Global Accord: Environmental Challenges and International Rersponses.* Cambridge, MA, MIT Press: 1993.

sustainability issues. In 1997, NGOs delivered more assistance to developing countries than the entire United Nations system. In a real sense, then, the authority of the national state in industrial ecology and environmental policy, while still dominant, is increasingly devolving to both NGOs and, of course, transnationals, with implications for local, national, and international legal structures that are not yet apparent.

12.1 FUNDAMENTAL LEGAL ISSUES

It is not that legal treatment of environmental issues is new. In 1306, for example, London adopted an ordinance limiting coal combustion because of the degradation of local air quality. Such laws proliferated as industrialization created new impacts. For instance, production of sodium carbonate (soda) by the LaBlanc process, patented in 1791, resulted in substantial emissions of hydrogen chloride, which led the English Parliament to pass the Alkali Act of 1863, which required manufacturers to absorb the acid in special towers.

Rather, what is new is the understanding of environmental issues as integral to human economic activity, not simply overhead. This implies concomitant changes in law and regulation dealing with the environment as existing environmental regulatory regimes generally reflecting an end-of-pipe, command-and-control approach, are augmented, and, more broadly, as environmental dimensions of non-environmental legal structures are identified.

Organizational implications are similarly profound. If environment is only overhead, a free-standing governmental entity, not significantly linked to any other, is adequate to meet all policy requirements. This is reflected organizationally in the assignment of responsibility for all environmental matters to one specialized organization, such as the United States Environmental Protection Agency.

If, however, environmental issues are strategic to society, then they must be incorporated, as appropriate, into virtually all other activities of society. Trade, security, procurement, consumer protection—all must be modified to reflect the integration of environmental considerations into them. This process, which may be complex enough in substantive terms, is frequently complicated by the fact that different groups of stakeholders, with different values and worldviews, must accommodate each other. Thus, in the United States for example, it is no longer adequate to have only the EPA making environmental policy; organizations such as the Departments of State, Defense, Commerce, and Energy, NASA, the Treasury, and even the Central Intelligence Agency find themselves increasingly involved in the environmental aspects of their core missions. Initial contacts are frequently somewhat contentious, as in the negotiations dealing with the appropriate role of environmental issues (and environmental groups) in the World Trade Organization and North American Free Trade Agreement systems.

In all these cases, it is not that the initial policy rationale behind each legal structure—free trade for mutual benefit, preventing vendor fraud of government procurement organizations, or whatever—is wrong. Rather, it is that the legal structure implementing that policy did not reflect any concern for environmental issues when it was initially created. Thus, a system that may be perfectly adequate for its original purpose may have unintended negative environmental impacts. Reducing these does not mean the original purpose is superseded by environmental considerations, but that the original structure is modified so as to integrate both sets of policy concerns.

Consider the example of the need to modify the substantive standards governing military procurement processes in the U.S., known as Military Standards (MILSTDs) and Military Specifications (MILSPEC), in light of the environmental requirement to reduce emissions of chlorofluorocarbons (CFCs), which contribute to stratospheric ozone depletion. Such MILSPECs and MILSTDs cover virtually all military process and product technology, and are an integral part of an extremely complex process. In fact, the single biggest barrier for the American electronics industry in its efforts to stop using CFCs in manufacturing pursuant to the international agreement, the Montreal Protocol, was not finding substitute technologies or capital, but MILSPEC and MILSTD (the military being a large customer for many American Electronics firms). Meeting the new, environmentally driven technology requirements could not, of course, be done at the expense of the original purpose of the specifications: to ensure high quality, readily available, robust military systems. Instead, new specifications that maintained or, in many cases, improved product and system performance without the use of CFCs were developed. The policy goals were integrated with the importance of both recognized—but the process was a long one, and the potential for conflict as integration proceeded was always present.

Similarly, the environmental desirability of reusing components does not mean that consumer protection laws that discourage sale of used equipment as new, and thus reduce the possibility of consumer fraud, are wrong. Rather, it means that ways must be found to integrate the purposes of both: Encourage the use of used components while continuing to guard against vendor fraud.

The process of integrating environmental with other policy structures is just beginning. The discussions, conflicts, and resolutions to which this process gives rise are shaped in part by a relatively limited number of fundamental legal issues, which are discussed in the following sections.

12.1.1   Intragenerational Equity

The distribution of wealth and power both among national states, and between the elites and the marginalized populations within individual national states, is one of the thematic foundations of political science. It is a highly ideological and contentious arena, in which the interplay between law and culture is particularly charged. The assumption of many formulations of sustainable development, which contemplate reasonably equal qualities of life for all people, implies a substantial shift in resources between rich and poor nations, as well as within nations. Whether and how this should be accomplished are the subjects of extensive debate. In part, for example, some of the opposition to the concept of sustainable development in the United States can be attributed to the perception that it is simply a device by which developing countries seek to redistribute income from the wealthier countries to themselves, and is thus simply a continuation of decades-old demands under new guise.

12.1.2   Intergenerational Equity

The very concept of sustainability across generations implies some representation of the interests of future generations in current disputes—such as storage of radioactive residues from electric power production, deforestation and reforestation, the extent of climate change mitigation efforts—that might affect those interests. While many legal and philosophic traditions recognize the general concepts of fairness between generations, and take some degree of responsibility for stewardship of the earth's resources, actual practices tend to be less sanguine.

There are three reasons for this. First, as noted above, the legal profession and its intellectual tools have been developed during a period of essentially unlimited growth of population and resources enabled by the Industrial Revolution. Although intragenerational resource conflicts were and are legion, this assumption negates the need for any allocation as between existing and future generations. After all, given few limits to growth in resource availability, future generations will, virtually by definition, have what they need.

More practically, perhaps, future generations are not viewed as having legal rights because such rights can only arise when there are identifiable interests, and it is almost impossible to identify either future individuals or their interests with sufficient specificity to involve them in adjudication of such interests. Who, for example, knows what resources will be critical to future technologies? Who can say for sure what their preferences will be? Assuming that some degree of intragenerational inequality still exists, whose interests will be represented: those of the elites? The disenfranchised? The practical problems involved with establishing such representation are apparent on a moment's reflection.

Finally, the scope of most environmental perturbations of concern are global or at least international. Moreover, in most cases the costs and benefits of both the impact itself, and of potential mitigation activities, are differentially distributed among national states and stakeholder groups across different temporal scales. As a result, achieving negotiated legal resolutions is extremely difficult, and enforcement mechanisms are non-existent. There are thus significant practical difficulties in evaluating intergenerational equity questions, much less establishing an institutional structure within which they can be resolved.

Notwithstanding these difficulties, it is, in principle, possible to develop an outline of an international system of legal obligations and duties that can support the implementation of intergenerational equity. Edith Brown Weiss, for example, proposes three basic principles of intergenerational equity:

1. The "conservation of options": Each generation should conserve natural and cultural resources so that the options available to future generations are no less than those enjoyed by the current generation.

2. The "conservation of quality": Each generation should conserve the quality of the planet in a like or better state than that in which it was received.

3. The "conservation of access": Each generation should provide equitable access to this global legacy to all its members, and conserve such equitable access for future generations.

In line with the general thrust of sustainable development theory, the third principle links intergenerational and intragenerational equity considerations, and promulgates the latter through time.

These principles then become embodied in duties: The duty to conserve resources, to ensure equitable use of available resources, to avoid significant adverse impacts and prevent disasters, to minimize damage and provide emergency assistance, and to compensate for environmental harm. While these duties are extremely general, in principle they could be defined through the development of interpretive case law applying them to specific factual situations (this is precisely how American and English Common Law is evolved).

12.1.3   Flexibility of Legal Tools

Because legal systems in many societies tend to be important components of social structure, they are usually conservative and relatively inflexible. One does not want a situation, for example, in which inheritance laws are changed every six months: The passage of property between generations, the assurance of free title, and preservation of family harmony and expectations all argue for a reasonably stable inheritance system.

There is a price, of course, for this inflexibility, which is paid in terms of inability to adjust to changing situations. Where such change is limited and foreseeable, as it may be with inheritance issues, this inflexibility is relatively unimportant. Where, however, change is rapid and fundamental, as it currently is with environmental issues, such inflexibility can lead to substantial inefficiency. This is especially true as environmental regulations extend from command-and-control end-of-pipe requirements—which may be expensive, but don't really impact choices of materials, manufacturing technologies, product design, or customer choice—to pollution prevention and product regulation. For example, an overly conservative requirement for scrubbers might cause manufacturing plants to spend marginally more than they should on such technologies. An overly conservative process or product standard, however—or, worse yet, one that subsequent data demonstrates was environmentally inappropriate—skews manufacturing, product design, and consumption patterns for a long time, as it becomes embedded in industrial systems, and usually cannot simply be substituted out.

The initial inflexibility built into such systems procedurally is augmented by the well-known tendency of regulation to create and nurture interest groups that benefit from continued regulation, and can form a significant barrier to subsequent regulatory rationalization. For example, reform of U.S. hazardous waste law is substantially impeded by a number of interest groups, including the legal and engineering firms that benefit from them, the firms that produce government-approved technologies and waste management services, and, more subtly, the environmental groups that use the fear of hazardous waste as a membership recruitment and fund-raising device.

Industrial ecology policy is an area where it is highly desirable to use flexible regulatory tools that can be adjusted as new data, technological options, and industrial practices warrant. The Netherlands, for example, uses "covenants" negotiated between industrial groups and the government, similar to contracts, to establish industrial performance requirements (see Chapter 17). Such covenants can be modified through negotiation as conditions warrant, however, without the tedious and contentious legal process usually required to modify regulations or statutes. Such a process also discourages the formation of interest groups that ossify an already rigid process.

In order for such a process to work, several critical requirements must be met:

- *There must be adequate transparency to the policy development process.* All stakeholders with a legitimate interest in the outcome should be represented as the regulations and implementation plan are developed.
- *There must be performance validation mechanisms.* Whether it is deployment of sensor systems or data reporting requirements, or implementation of third-party inspections, there must be some objective means by which performance can be validated by the government and interested members of the public.
- *There must be long-term metrics or standards in place.* If a policy or program is to be subject to continued modification, there has to be some goal against which potential changes can be judged and either accepted or rejected. These need not be quantified or specific, but they must constitute an adequate filter for proposed modifications to the initial agreement.
- *The process, including conditions under which both parties agree to consider modifications to the original agreement, must be explicit.* This is necessary to ensure that unnecessary conflict is not generated between the government and the covenanting industry group, as well as to maintain public transparency, and thus political viability, for the process.
- *The legal structure of the national state must allow for government negotiation and enforcement of such quasi-private agreement.* Sweden, for example, is well aware of the benefits of such flexibility, but the Swedish Government, under its Constitution, cannot negotiate such agreements.

12.1.4 Regulatory Management Structure

Traditional command-and-control regulatory mechanisms have been applied with significant success to easily visible environmental problems, particularly those involving point-source emissions such as manufacturing facilities. Unthinking extension of such simple regulatory tools to far more complex situations, which arise with increasing regularity as environmental issues move from overhead to strategic for firms and society as a whole, however, can frequently be both environmentally and economically costly. Rather than continuing to rely primarily on centralized command-and-control, a more sophisticated environmental management system, which recognizes the complex nature of the systems at issue, must be evolved (see Chapter 9). This, in turn, requires more attention to balance along two dimensions of regulation.

12.1.4.1 Boundary Conditions versus Targeted Intervention

For different environmental and technological issues, policy makers must decide whether it is more appropriate to impose targeted, specific requirements, or to establish broader boundaries on behavior that encourage appropriate system evolution over time. This, in turn, requires an evaluation of the context of the proposed regulation or policy.

1. Is the environmental impact significant in spatial terms and potential damage, and difficult or impossible to reverse? If so, this might justify targeted intervention against the activity causing the impact.

2. Is the environmental impact the result of diffuse activities throughout the economy, with no easy technological fix? If so, this might justify changing the boundary conditions, rather than targeted command-and-control intervention: Regulators are highly unlikely to have adequate information to define and implement efficient mitigative measures themselves.

3. Are potential ancillary impacts of the proposed policy easy to identify and control? If so, targeted intervention is more easily accomplished and the costs and benefits easier to evaluate.

For example, in the case of removing lead from gasoline, where the impacts on human and ecosystem health are both significant and relatively obvious and the ancillary effects relatively easy to determine, targeted intervention in the relatively onerous form of a ban is a justifiable policy. In the case of global climate change, however, it is hard to identify any form of targeted intervention that could both significantly reduce anthropogenic forcing, and be economically and environmentally efficient. In the latter case, providing boundary conditions in the form of energy taxes or fees, combined where possible with sector-specific initiatives such as energy efficiency labels for electronics equipment, may make a lot more sense.

12.1.4.2 Decentralized Mechanisms versus Centralized Micromanagement

A similar set of issues arises when choosing between reliance on decentralized mechanisms, principally the information and operational efficiency of markets, as opposed to centralized micromanagement. Conscious reliance on decentralized mechanisms generally involves setting boundary conditions and letting the market respond, while targeted intervention is the tool of centralized micromanagement. The boundary between these two approaches need not, however, be clear in practice. For example, banning chlorofluorocarbons because of their effect on the stratospheric ozone layer is a targeted intervention against a specific class of materials, but did not in most cases involve centralized micromanagement. Indeed, the process of technological substitution was encouraged in the U.S. by implementing a tax on CFCs as well, combining command-and-control and boundary setting mechanisms fairly successfully in an integrated policy. Firms were explicitly encouraged by regulators to develop their own technological alternatives to CFC use, and governments, industrial groups, and NGOs worked rather well together to evaluate alternative technologies.

The important question here is really to what extent a given policy should, or even must, rely on market forces. While the specifics of each issue are important, it can be stated as a general principle that, overall, both environmental and economic efficiency benefit if market mechanisms are used to the maximum extent possible.

In part, this is because once the obvious environmental insults to air and water and human health are addressed, the remaining ones tend to manifest themselves heterogeneously over time and space. So do many dimensions of the relevant human and natural systems, such as resource and sink availability; structure, resiliency, value, and structure of biological communities; infrastructure availability; cultural considerations; and technological capabilities. Such scalar variability tends to argue for caution in relying on centralized regulatory mechanisms in many cases. Increasingly, then, it becomes advisable for regulatory management and control functions to be distributed so as to reflect the inherent complexity and structure of the issues being addressed.

Relying on market mechanisms has many advantages under such circumstances, but there are a number of caveats as well. Among the most obvious are those situations in which, for a variety of reasons, prices don't reflect the cost or value (in both an economic and moral sense) of important components of the system. Under these circumstances, the political process must be relied upon to temper naive recourse to market mechanisms.

12.1.5   Determining Appropriate Jurisdictional Level

Political jurisdictions are creations of human culture and history, and there is no a priori reason why their boundaries should reflect underlying natural systems. It is thus no surprise that many problematic environmental perturbations are not coextensive with political boundaries. Emission of acid rain precursors in the United States or China cause acid rain in Canada or Japan; watershed degradation involving different national states generates enormous legal and political conflict.

More subtly, industrial or consumer behavior may not be geographically or jurisdictionally co-located with the environmental perturbation to which it contributes. Thus, for example, much of the environmental impact of the economic activity of developed national states is already embedded in the products or materials they import. Given the usual condition of prices that do not include all relevant social costs, these embedded environmental costs will thus be virtually invisible to policy makers and consumers. Because the reach of the national state usually does not extend beyond its borders in such cases—indeed, in many cases it cannot because of relevant international requirements such as trade law—this separation of behavior from impact can make identification and management of such situations difficult.

It is possible, however, to reduce the distorting effects of jurisdictional boundaries that are not coextensive with the scope of the issues under consideration. This can be achieved both horizontally (by harmonization of policy across jurisdictions of equal legal status) and vertically (by integration of policy structures up jurisdictional hierarchies).

12.1.5.1 Policy Harmonization

Harmonization of regulatory management structures across jurisdictions of equal status is desirable in many cases for several reasons. First, commerce is increasingly regional and global in scope, so the imposition of substantively different constraints on economic activity at small geographic scales can cause significant economic dislocation. This is of some concern because in many cases the intent may be less to protect the environment, and more to impose trade barriers that protect local business. There are a number of cases involving beverage packaging in the European Union, for example, where requirements imposed for environmental reasons—such as mandated use of refillable glass bottles, which in practice requires local bottling operations—have such obvious trade implications that motives become questionable.

Another important reason is the problem of exported risk. Where a particular substance or activity is banned or stringently regulated in one area, yet demand for the regulated item remains strong, it will frequently simply be displaced to a jurisdiction where it is less regulated. Thus, for example, environmental dimensions of mining activities are heavily regulated in the United States and Canada, with the result that much new mining activity has now shifted to developing countries. Harmonization of international standards of environmental performance can reduce this problem. The Basel Convention, for example, attempts to reduce export of risk in the form of hazardous wastes, although the potential of such structures to constrict environmentally desirable material recycling behavior is not well understood by many regulators.

There is an important caveat to harmonization efforts, however: They are appropriate where the environmental perturbation of concern is homogeneous over a broad area, but not where heterogeneous phenomona are involved, thus requiring differing local standards. For example, the Montreal Protocol addressed the emission of ozone depleting substances at the global scale, which, given the behavior of the emitted substances and the global averaged scale of the resultant impact, was appropriate. Environmental regulations affecting land use, which is a quintessential local issue, however, are best tailored by local jurisdictions to their requirements. An example of inappropriate harmonization is Superfund, the hazardous waste cleanup law in the United States, in that it has applied the same cleanup standards to property across the country regardless of its intended use with a strict, joint-and-several liability scheme. The result is that many previous industrial sites are not "recycled" to new industrial uses because of the possibility of inheriting liability for any contamination that may exist, and industries turn to so-called "greenfield" development—that is, they preferentially clear new land rather than reuse already developed land. The results are disinvestment in urban industrial sites and unnecessary conversion of agricultural and less developed land to industrial uses, all driven by environmental law. Partially as a result of pressure from states and localities, this policy is changing, albeit slowly.

A particular problem for harmonization arises when public opinion results in the passage of "symbolic legislation," which is intended to demonstrate concern, but not to be seriously implemented. For example, in addition to the Montreal Protocol and implementing national legislation, a number of municipalities passed regulations addressing the same phenomenon, in some cases adopting different standards, time-lines, and technology requirements than national or international agreements. The impact on the underlying phenomenon of such "symbolic legislation," which in most cases is not enforced, is usually small to nonexistent, but substantial economic ineffi-ciencies can nonetheless be generated. Moreover, it is a useful general principle that laws which make customary and routine behavior illegal are to be avoided, and such symbolic legislation, even if not enforced, results in inadvertent, sometimes virtually unavoidable, illegal behavior. For example, one U.S. state outlawed the use of a partic-ular type of packaging made using CFCs that was legal everywhere else in the world. No effort was made to enforce this law, and discussions with state politicians indicated it was passed to reflect the concern of a number of local groups with stratospheric ozone depletion—in other words, as symbolic legislation. One firm immediately, and at some cost, terminated shipments of its products only to that state. Most others, many of whom were not even aware of the existence of the legislation, violated it.

Even where the physical aspects of the phenomenon are global and local condi-tions are relatively homogeneous, harmonization may not be appropriate where local values differ. In particular, risk is a somewhat subjective phenomenon (see Chapter 10), and thus different cultures may act differently, albeit rationally, because of differ-ent perceptions of risk. For example, some Northern European countries such as Sweden are considering banning polyvinyl chloride (PVC), a common plastic, because of their concern about chlorinated compounds. Others, even in the European Union, using precisely the same data believe that the risk posed by PVC is not significant com-pared to its benefits.

Harmonization issues thus reflect the underlying phenomenon of an economic system that is at the same time globalizing and becoming more heterogeneous. Both global and local standards, in many cases at least potentially contradictory, are prolifer-ating, with impacts on process and product design and materials and technology choice. The goal of policy should not be to impose global harmonization unthinkingly, nor to allow anarchistic proliferation of local standards and practices where appropri-ate. Rather, more sophisticated policies aligning jurisdictional boundaries with the scale of the underlying phenomena should seek a reasoned balance negotiated in a transparent, open manner.

12.1.5.2   Integrating Policy Hierarchies

Just as policies should be integrated horizontally to the appropriate extent, they should be integrated vertically as well. Unlike harmonization, however, this require-ment will generally not result in adoption of the same policy. Rather, policies at each jurisdictional level, while addressing the specifics of the concern at that level, should consider their relevant implications for all levels: at the least, policies should not create unnecessary conflicts among jurisdictional levels. Thus, for example, national laws regarding visual monitoring of fuel storage tanks to minimize leaks into the environment should not conflict with pre-existing local fire regulations requiring that gasoline tanks be stored underground for safety reasons. Similarly, policies regarding industrial production in an upstream riparian national state should minimize not only the direct impacts of production, but also its effects on downstream areas. Such export of risk is not a substitute for risk reduction, and should be avoided because it encourages the generation of externalities (not to mention conflict; see Chapter 18) when the system is viewed as a whole.

In practice, integrating policies vertically can be quite difficult unless the situation being addressed is relatively simple. As the "symbolic legislation" example given above demonstrates, even in the case of stratospheric ozone depletion resulting from emission of CFCs, where both the behavior of the physical and chemical system and the industrial activity resulting in the emissions were relatively simple, vertical policy integration was difficult. In the case of more complex perturbations, such as global climate change or anthropogenic alterations of the nitrogen cycle, policy integration becomes quite difficult. In practice, integration may be best achieved not by trying to blend policies on the legal side, but by developing a set of metrics that can cascade relatively gracefully through policy levels.

12.2   SPECIFIC LEGAL ISSUES

A survey of some of the specific areas where existing legal structures come into conflict with environment as it becomes strategic rather than overhead for society will lend concreteness to the concept, as well as illustrate the breadth of the legal issues involved. Inevitably, such a discussion must reflect particulars of the legal systems and countries involved, but the examples are useful nonetheless, and the principles illustrated can easily be extended to different legal systems. Accordingly, a number of examples are briefly discussed below. Chapters 16 and 18 provide more extensive case studies in two areas: the first is a conceptual study of the legal definition of the private firm, while the latter discusses the definition and implementation of environmental security policy, which combines national security and environmental considerations.

12.2.1   Trade and Environment

The potential conflict between trade and environment is apparent. Trade policy as reflected in international agreements, such as the North American Free Trade Agreement (NAFTA) and organizations such as the World Trade Organization (WTO), by and large seeks to facilitate the free transfer of goods and services among national states. The environmental community, on the other hand, seeks to control trade in environmentally unacceptable goods and, in some cases, to impose restrictions on the means by which national states produce goods and services internally. The situation is complicated by the fact that environmental arguments are sometimes used to justify restrictions on trade that are, in fact, "protectionist"; that is, their purpose is not so much to protect the environment, as to provide protection for domestic economic interests against otherwise lawful foreign competition.

In some cases, this potential conflict between trade and environment becomes real; free trade values actually are opposed to environmental interests. For example, some European countries have imposed requirements that beverages be sold in returnable glass containers. The environmental purpose is to reduce the amount of waste produced by plastic or paper containers that are discarded, and to encourage the reuse of containers as opposed to the recycling of the material from which they are made (the degree to which the latter is environmentally preferable, and under what conditions, remains somewhat unclear). On the other hand, because of the weight of glass bottles, and the difficulty and expense of the reverse logistics system by which the bottles must be recovered and reused, such a requirement clearly favors local (domestic) bottling operations and beverage producers such as brewers.

Under such circumstances, several general principles have been established to resolve disputes:

- Any environmental restriction on trade must be neutral in application; that is, it cannot be applied explicitly only to foreign firms or products, and not to domestic products.

- The environmental benefit claimed must outweigh the detrimental impacts on trade. This obviously is a weighing test, fact-specific in application, intended to prevent the use of environmental rationales for protectionist purposes. For example, a requirement that foreign lead not be imported because the transportation of such a heavy material would cause carbon dioxide emissions and thus contribute to global climate change would not be accepted under trade law: The alleged environmental benefits of the requirement are minor compared to the obvious impact on trade.

- The specific restriction on trade imposed by the environmental requirement must be the least possible to achieve the desired environmental goal. In the above case, for example, it is obviously possible to control emissions of carbon dioxide much more effectively through other means that have far less impact on free trade.

- As a general matter, environmental policies that attempt to limit how a country produces an otherwise acceptable product are disfavored. This reflects not only the difficulty of imposing global environmental standards where technologies and values may differ greatly between the countries involved, but the general disinclination of international law to permit one country to dictate the internal policies and practices of another. This previously absolute primacy of the national state in international law is, however, eroding over time in many areas: Certain human rights, for example, are increasingly viewed as inherent to all people, and thus not subject to violation even by a national state's internal processes.

12.2.2 Consumer Protection Law

The general purpose of environmentally relevant consumer protection law is to encourage full disclosure of the properties of the product by the vendor, and thus avoid fraud. Basically, used products are not to be passed off as new ones. Thus, in part, consumer protection laws require that any product which is used, or contains used parts, be prominently labeled; the idea being that used products or parts are frequently inferior to new ones, and should not be foisted on unknowing consumers. Such a label, not surprisingly, significantly reduces the price that can be charged for most articles, and can hurt the trademark of the producer or vendor. This provision thus provides strong incentives against recycling used components, subassemblies, or products.

Reuse of products or parts, however, can provide clear environmental benefits, and should thus be encouraged by public policy. Moreover, in some instances—memory chips, for example—there is little difference between a new and a used part, and in many other cases a used or refurbished part is more than adequate for the use for which it is intended. In such cases, so the argument runs, consumer protection law inappropriately discriminates against environmentally desirable practices.

Although this conflict between the legitimate policy goals of consumer protection law and environmental protection law has yet to be resolved, there are several obvious possibilities. As a stopgap measure in the short term, the principle could be established that, so long as a product, component, or part meets all relevant specifications, it is immaterial whether it is used or not. In the longer term, the issue is one of consumer education: Customers have been acculturated to avoid used products, or to value them less, and will need to be educated about the benefits of consuming used products. This process can be assisted by internalizing the positive externalities of such informed consumer choice—in short, by passing along the savings of using refurbished products and components to the consumer.

Interestingly enough, the best solution may arise with the continued evolution of the "functionality economy," where customers increasingly purchase function or services, rather than the underlying physical product. This sounds somewhat daunting, but it is not: leasing of cars and office machines are current examples. In these cases, the customer doesn't care whether the product contains used parts or not: They are contractually assured of performance in any event. Thus, for example, before it was broken up, the Bell System in the United States leased telephones to all subscribers. Although most phones were used or contained used parts, the user knew they would always have a working telephone, so this was immaterial to them. More recently, Xerox, which leases copying equipment to businesses, has even made its practice of refurbishing copiers and reusing parts and subassemblies a way to differentiate their products from those of competitors in the European market, where environmental image is important. Not only is Xerox selling document reproduction rather than copy machines, but it has turned this shift in business perspective into a marketing initiative. Environmental responsibility thus no longer conflicts with consumer protection law, but actually becomes a source of competitive advantage.

12.2.3   Government Procurement

In a similar effort to avoid vendor fraud, many governments have requirements embedded in their procurement regulations that actually prohibit the purchase of used products or products containing any used components. Incentives for this kind of procurement constraint may be less commendable as well: Such requirements can be used as trade barriers (as has occurred with Xerox and local competitors in Europe). The situation differs from that of consumer protection law slightly, however, as consumers presumptively have their own interests at heart in procuring the best product for the price, while government procurement offers the possibility of collusion between the purchasing entity and the vendor to buy inferior goods (and split the savings). Nonetheless, there does not appear to be any reason why the possible solutions applicable to the consumer protection situation cannot also apply here: purchase according to relevant specifications, or purchase functionality or service rather than product.

Government procurement practices are more important for their potential positive impacts on industrial behavior, however. End-use, individual consumers are a broad, unorganized group, and most consumers do not make purchases for environmental reasons (despite what polls frequently find; many people are aware that they should say they will purchase environmentally preferable products, but in practice they do not do so). Governments, on the other hand, have substantial buying power centralized in one organization, and thus can exercise significant control over a market; by some estimates, OECD governments control between 5 and 15 percent of all consumption. To the extent government procurement practices can be made environmentally preferable, therefore, they can exercise significant beneficial impacts on the performance of producers and vendors. This tool, frequently underutilized, is potentially a significant lever for environmental progress (and, if done properly, can avoid many of the inefficiencies associated with command-and-control). To put it in theoretical perspective, changes in government procurement are a way of internalizing environmental considerations that were previously externalities.

12.2.4   Government Standards and Specifications

Few people outside government and the industries that supply them recognize the pervasiveness and power of government product and process standards and specifications, especially those associated with military procurement. These standards and specifications control a substantial amount of the design of many products and the processes by which they are made, and, in many cases, predate any concern with the environment. They thus frequently embed environmentally problematic requirements within the economic system, and do so in a way that is invisible to most people. This effect is magnified because many private purchasers, recognizing the concern for performance and quality built into such specification systems, reference them directly or indirectly in their own procurement documents.

Thus, for example, the single biggest barrier to the American electronics industry's efforts to stop using chlorofluorocarbons (CFCs), which were contributing to the breakdown of the stratospheric ozone layer, were military specifications and military standards (known as MILSPEC and MILSTD). Moreover, because of the tens of thousands of references to such requirements in myriads of procurement contracts and subcontracts, an enormous amount of work had to be done simply to change the welter of legal restrictions on using anything but CFCs. Given all this, the technology involved in developing substitutes, and "proving them in"—demonstrating their efficacy—was relatively trivial by comparison.

The obvious solution is to change such standards and specifications from being barriers to environmental and economic efficiency to supporting such efficiency, while maintaining at all times the underlying requirement of guaranteed performance that generated the existence of such standards and specifications in the first place. This is easier said than done, however; the pervasiveness, complexity, and sheer volume of such requirements are significant barriers to improvement. Some current initiatives, such as the shift from U.S. military procurement to civilian products and parts where they are adequate, should help to alleviate such barriers somewhat. Nonetheless, the potential power of MILSPEC and MILSTD, and government standards generally for generating environmental and economic efficiency has yet to be fully explored.

12.2.5 Antitrust

As in the case of trade and environment, there are some fundamental issues regarding the relationship between antitrust and environmental policies. Antitrust seeks to maintain the competitiveness of markets by limiting the market power of firms, which generally means limiting their scope and scale. Based on standard neoclassic economics, antitrust generally seeks a marketplace where there are so many buyers and sellers that no single entity has the power to extract rents from the market. Many environmental initiatives, such as post consumer product takeback, however, seek to do the exact opposite: to expand the scope and scale of the firm so that it is responsible for the environmental impact of its product from material selection through consumer use to takeback and recycling or refurbishment. The one seeks an atomistic market with no central control; the latter seeks to extend the control of firms in the interest of internalizing to them the costs of negative environmental externalities (and benefits of positive externalities).

The dichotomy between antitrust and environmental policies is exacerbated by the question of technological evolution. By and large, technological evolution is most rapid in competitive markets with low barriers to the introduction of new technologies. Such market structures may well be fostered by traditional antitrust policies. On the other hand, if firms are to implement environmentally preferable practices across the life cycle of their product, they will generally have to develop means of linking the technologies used at various points in the product life cycle. Thus, for example, the technologies used to disassemble the product after the consumer is through with it need to be considered in the initial design of the product (a process called by designers, reasonably enough, "Design for Disassembly"). Linking technologies in such a way creates a more complex, co-evolved, technological system, and reduces the ability to evolve any part of that system rapidly. It "locks in" technology choices.

Thus, on the one hand industrial ecology indicates that rapid evolution of environmentally and economically more efficient technologies is critical to moving toward sustainability in the short term, but, on the other hand, encourages the development of systems that reduce the potential for such evolution. The solution to this dilemma—to understand which structure is economically and environmentally better under what conditions—requires an analytical sophistication and an integration of antitrust and environmental law that does not yet exist.

12.2.6 Existing Environmental Law

As a final example, it is useful to consider the somewhat ironic situation that existing environmental law in some cases is becoming a barrier to continued improvement in environmental and economic efficiency. On a moment's reflection, however, this is not surprising: Like other policy systems, existing environmental regulations were predicated on the implicit assumption of environment as overhead, and thus, like any other policy system, will generate conflicts as environment becomes strategic.

It is important to remember that existing environmental laws in developed countries have produced substantial environmental benefits: The question is not doing without, but rather recognizing the situations that require a more sophisticated approach given a more sophisticated understanding of environmental perturbations. It is also important to note that, in many cases, the problems with existing law arise because they have generated powerful interests groups that now have a vested interest in maintaining the status quo, even if it is not environmentally (or economically) preferable. The difficulty is not in the law or regulation itself, therefore, but in the political process that precludes desirable change in light of new circumstances.

The complex of "waste" laws that have built up at the national and international level will serve to illustrate this problem. A common characteristic of such laws is that the definition of "waste" is a trigger: If a material is not "waste," it is not regulated, but as soon as it becomes "waste" it is subject to heavy regulatory burdens. Moreover, the "waste" designation is taken to reflect to a large degree the inherent properties of the material, leading to, for example, provisions in U.S. law to the effect that, once a material is a "waste," it will always be a "waste." Such provisions, which to an engineer or materials scientist seem absurd, generally are intended to avoid fraudulent disposal of hazardous residues, or their export to less stringent jurisdictions under the guise of shipping commercial materials.

The purpose of such requirements, to assure safe handling of hazardous materials, is laudatory. In practice, however, such laws tend to become problematic in two ways. For one, they regulate materials that could be recycled far more heavily than virgin materials, thus distorting material markets to favor virgin rather than recycled materials. Moreover, they have a subtle psychological effect that becomes embedded in society: People who study and work with waste are separate from those who work with production processes, technologies, and product design. Two different communities evolve, and the lack of communication between the two further exacerbates the difficulty of integrating end-of-life, post-consumer product and material management with initial product design and material choice. "Waste" laws create communities and incentives that embed waste production into the economy.

A related difficulty with such an approach can be seen by considering the concept of manufacturing implicit in such laws as the Resource Conservation and Recovery Act (RCRA) in the United States. The manufacturing paradigm implicit in RCRA, which defines hazardous materials that fall out of the manufacturing process as "waste" rather than residuals, presupposes a linear, Type I system. It assumes that the manufacturing system, taken as a whole, takes materials in at one end, puts them into products or processes, and puts it out the other end as either a product destined for consumers or "waste." In assuming this, and regulating such "wastes" heavily, it in fact provides strong incentives to maintain a linear manufacturing system, and prevents the evolution towards a more cyclic system.

It is not that the concern about inappropriate material management inherent in such laws is wrong; indeed, the number of material recycling operations that have caused environmental problems in the past is adequate testimony to that. Rather, the problem is the inability to evolve the regulatory system to the desired state—similar regulation for materials used in ways that cause similar risks or environmental problems, based on the nature of the activity or material, not its regulatory status.

REFERENCES

Cairncross, F. *Costing the Earth.* Boston, Harvard Business School Press: 1992.

Daly, H.E. and J.B. Cobb, Jr. *For the Common Good.* Boston, Beacon Press: 1989.

Graedel, T.E. and B.R. Allenby. *Industrial Ecology.* Upper Saddle River, NJ, Prentice Hall: 1995.

Weiss, E.B. *In Fairness to Future Generations.* Dobbs Ferry, NY, Transnational Publishers, Inc.: 1989.

World Commission on Environment and Development. *Our Common Future.* Oxford, Oxford University Press: 1987.

EXERCISES

1. You are the Environment Minister for an enlightened developed country. Your Prime Minister has called you into her office and expressed some dissatisfaction with your existing environmental statutes and regulations, which regulate releases to air, water, and land in the traditional command-and-control, end-of-pipe manner. She gives you the responsibility of developing a set of policies that will lead to sustainability in one generation (25 years).

    a. What policies will you recommend, and why?

    b. What changes, if any, will you make in your current regulatory structure?

    c. Does it matter if you are a large country constituting a significant international market or a small country that imports most of your products and exports most of what you manufacture? Why or why not?

    d. What will you tell your Prime Minister about the possibility of achieving her goal?

2. You are the Attorney General of a progressive, developed country. Pursuant to instructions from your Prime Minister, you are determined to represent the interests of future generations in your current legal processes.

    a. How will you recommend that this be done?

    b. How does your answer change if instead of being the Attorney General of a country, you are the Chief Legal Counsel of a private firm?

3. Do you think it is possible to achieve intergenerational equity without achieving intragenerational equity? Defend your answer.

4. Symbolic legislation, which is passed to reflect public concern but not intended to be implemented, would appear to impose costs without commensurate benefit, yet it continues to be passed.

   a. What advantage do you see to passing such legislation?

   b. Some critics have said that even some environmental legislation, such as the Clean Water Act in the United States, is symbolic, in the sense that the legislative standards established in the legislation cannot be met, and that the implementing agency, EPA, is not provided with sufficient implementation resources in any event. Assume that this criticism is at least partially true. What are the benefits and costs of passing such legislation anyway?

Government Structure and Industrial Ecology Policy Formulation

13.1 OVERVIEW

In a survey text such as this, a complete discussion of the complex relationships between government structure, cultural systems, and policy formulation is obviously impossible. It is a subject deserving of lifetime study. Even if limited to those aspects of governmental organization that have particular relevance for industrial ecology policy formulation, it is a daunting task, for such policy systems will at the least include important elements of environmental policy, technology policy, research and development policy, national security policy, and economic policy. Nonetheless, the next few chapters will provide an introduction to this complex and multidimensional policy field. This chapter begins with a few general comments about industrial ecology policy formulation, and then identifies several dimensions of structure and organization that are significant in terms of industrial ecology policy.

13.2 POLICY LIFE CYCLE

As Marion Chertow of Yale University has pointed out, it is important to remember that, like environmentally appropriate products, processes, and operations, rational policies have a life cycle as well. Although, as with products, there are a number of ways to break the policy life cycle down, a common formulation includes six stages. The first, *initiation*, involves the recognition of a need for a policy at all; in systems terms, it is the beginning of the internal response to a perceived change in external environment. The second, *estimation*, is the assessment of options for response: What policies might respond to the change in external conditions? The third, *selection*, is the choice of policy response; this is usually a complex process, involving not only the factual conditions involved, but economic, political, cultural, ideological, and even theological dimensions (think of population control). The fourth, *implementation*, is self-explanatory, but, like the selection stage, is often complex and problematic in practice. The fifth, *evaluation*, is the policy assessment stage: Is the policy producing the desired outcomes, and what unanticipated outcomes are occurring that might justify changes in policy, and what might those changes be (looping back to the beginning of the process)? The sixth, obviously, is *termination*: When the policy has accomplished its desired goals, or when external conditions shift such that the policy is no longer needed or is contraindicated (e.g., new data are discovered invalidating the technical assumptions that underlie the policy), the policy should be terminated. Anyone with any experience knows that this last stage is always problematic, because policies inevitably generate stakeholder groups that benefit from them and will fight vigorously to retain them.

Interestingly enough, this formulation of the policy life cycle, if it is done properly, is similar to the Total Quality Management processes that have become so popular in industry. The essential dynamic of both is to plan, implement, and then evaluate and begin the process over again, an iterative process, which, if done properly, results in continuing incremental improvement in performance. There are two caveats, the first being the common observation that, if it were only this easy, industry and the world would be a much better place.

The second caveat is more subtle. This conceptual framework supports incremental improvement but does not support discontinuous change. Given the improvements in environmental efficiency which many experts predict is required if sustainability is to be approached (Figure 13.1), incremental improvements are necessary but highly unlikely to be sufficient in themselves. Moreover, few policies are subject to the final two stages of the life cycle, evaluation and termination, until and unless they become

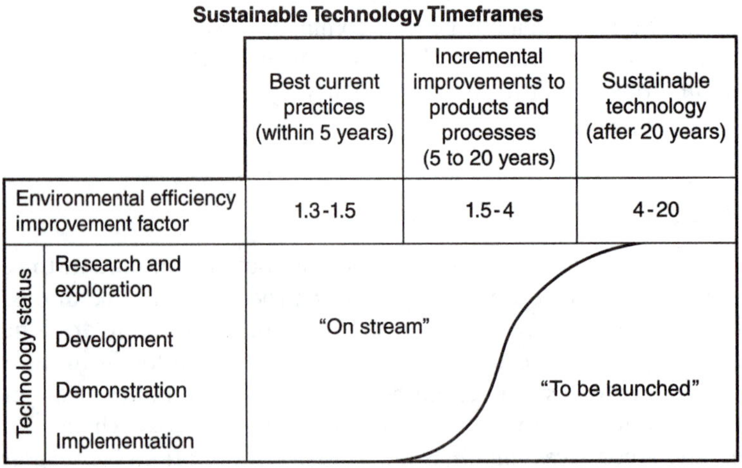

Sustainable Technology Timeframes

	Best current practices (within 5 years)	Incremental improvements to products and processes (5 to 20 years)	Sustainable technology (after 20 years)
Environmental efficiency improvement factor	1.3-1.5	1.5-4	4-20

FIGURE 13.1 Sustainable Technology Timeframes. Dutch experts estimate that truly sustainable technologies must provide improvements in environmental efficiency by factors of 4 to 20 over existing technologies. Such improvements cannot be obtained simply from incremental improvement to existing technological systems.

clearly dysfunctional. Under many circumstances, this is a limited problem, because the external conditions, which they were designed to respond to, change relatively slowly and predictably. In the case of industrial ecology and sustainability, however, where the economic and natural systems involved are changing rapidly and potentially discontinuously (as is our knowledge of them), it is as important to sunset outmoded policies as it is to generate important new ones. Thinking of policies in terms of life cycles makes this clear, if no easier to accomplish.

Additionally, the lesson of the old story about never wanting to eat at any restaurant where you have worked in the kitchen (because behind the good food lies a messy process) should be remembered. The political process by which policies are actually made, whether statutory or regulatory, is messy and confused as well. The policy life cycle concept is useful, but it is only a schematic; the process must be experienced for its full richness to be understood.

13.3   TEMPORAL AND GEOGRAPHIC SCALE CONSIDERATIONS

Any effort to develop industrial ecology policy must take cognizance of the fact that policies are generally short term responses on the part of a political system, and are inherently limited to the implicit or explicit scale of the jurisdiction involved. In some sense, the policy development system is thus similar to the psychological horizon of individuals (recall Figure 4.3); it has limits beyond which its effective function diminishes rapidly.

Accordingly, there is a fundamental problem in any effort to create policies designed for periods of significant change, which can be summed up as "getting ahead of society's headlights." At any given point in time, the psychological flexibility of most people, and political institutions, is limited. It is possible to propose change, but the proposed change must integrate gracefully with existing institutional structures, and, in general, cannot be discontinuous. Otherwise, valid though any proposals may be, they will not be accepted. Moreover, the greater the proposed change, the more threatening it will be to individuals, institutions, and stakeholders familiar with, and benefiting from, the status quo; another factor which reduces psychological and political palatability. Thus, for example, it is reasonably acceptable for environmental agencies to adopt policies favoring pollution prevention: even though such policies mark a fundamental shift from regulating wastes to regulating processes, that shift is an incremental enough step to be acceptable. It is generally less workable, however, for environmental agencies to adopt policies controlling the way people use their automobiles, even if it would achieve demonstrably greater environmental benefits. Such a change in perceived role is too great for rapid acceptance.

There is thus a difficult policy dilemma. The technical and policy experts may recognize that fundamental change is both necessary and inevitable if sustainability is to be achieved at all gracefully (that is, without substantial risk of economic or cultural conflict, or even increases in human mortality). They also recognize, however, that if they attempt to stretch their policies that far, at least initially and in the absence of crises, they will generate substantial opposition and have little chance of success.

Consider, for example, the likely possibility that a world with 8 to 10 billion people, all with a quality of life similar to that now enjoyed in developed countries, would most probably require technologies that produced quality of life at substantially reduced environmental impact (that is, with greater environmental efficiency). Figure 13.1, based on work done in the Netherlands, illustrates such an evolution toward "sustainable technology," defined as technology that provides significant enhancement of environmental efficiency (by a factor of, say, 4 to 20). Under these circumstances, policies that can affect technologies within 5 years, such as pollution prevention requirements, are reasonably achievable politically, and can be targeted with a degree of certainty. Identifying and implementing policies to support development of "sustainable technologies"—say, technologies and associated culture patterns that can, in fact, cut environmental impact per unit economic activity by an order of magnitude—is, however, far more difficult. There are a number of contributory reasons, such as the short time horizons of humans and their political systems, and the virtual impossibility of predicting the specifics of economic and technological change, especially in today's rapidly evolving global economy. Perhaps the most important reason, however, is that both people and institutions are resistant to change. This has several implications.

13.3.1 Increasing Flexibility and Generality with Time

Policies designed to respond to short term perturbations, and interventions in simple systems, can be targeted, inflexible, and rely on command-and-control mechanisms (which implicitly assume full knowledge of the system). The longer the time horizon of the policy, the more complex and uncertain the environment within which it will act, and the more flexible it must be. A policy designed to limit aqueous emissions of heavy metals from secondary smelters can be implemented in the short term, and in many cases will rely on command-and-control mechanisms. An effort to implement a policy intended to encourage sustainable patterns of use of heavy metals, on the other hand, is limited by uncertainties of all kinds: what really is sustainable; what production, consumption, recycling, and alternative technologies and materials may be available in the future; what economic and employment tradeoffs may exist. Initial policies may certainly be suggested—such as increasing fees on virgin material or reduction of unnecessary regulatory costs imposed on recycled metals—but their effect will be difficult to predict, and adjustments are likely to be required with experience.

Accordingly, the longer the time horizon of the policy, the more flexible and, at least initially, general it should be. In this sense, the covenant system of The Netherlands, where the government signs contracts with industry for enforceable, mutually agreeable, environmental improvements, is a preferable system to the rigid and difficult to change regulatory and statutory structure of, for example, the United States. An important component of such a policy structure is the development of metric and data gathering systems, which assure protection of the environment in a flexible policy context.

13.3.2 Reliance on Incentives

Short term policies intended to remedy obvious problems can rely on targeted command-and-control; policies intended to produce changes in long term behavior less so.

In the latter case, incentives that promote continued self-interested evolution by the affected community are more likely to produce results, with less political difficulty. In part, this is because the underlying scientific justification is likely to be more uncertain for longer term issues, which makes imposing costs to mitigate them less politically acceptable. Thus, for example, stratospheric ozone depletion is a global environmental perturbation, which, nonetheless, is being addressed by relatively short term policies because of the clarity of the immediate impact. Global climate change, on the other hand, is a longer term, less well characterized phenomenon and would require more fundamental adjustment of existing technological and economic systems to address. Accordingly, the political will to do something about it is correspondingly less. It is also a less appropriate issue for command-and-control because of its complexity and uncertainty; regulators cannot presume to understand in detail either the scientific or human (economic and cultural) systems involved. On the other hand, an incentive system that became stronger over time, and relied to the greatest possible extent on market mechanisms, would permit as efficient an evolution as possible to alternative, environmentally preferable, technologies.

13.3.3   Increasing Culture Change Dimension Over Time

Short term policies seldom must attempt to impose significant culture change; when they do, it is because of immediate crises that facilitate the necessary social acceptance. Policies intended for the long term, however, frequently involve cultural dimensions, which can be politically quite difficult. For example, it is likely that a sustainable economy will require a fundamental change in the currently quite strong linkage of material consumption and perceived quality of life. A fundamental cultural change of this nature will not happen quickly, and a policy that too obviously raises such a possibility from the beginning will also raise substantial opposition. This again highlights the importance of metric systems so that a series of short term, adjustable policies can be linked together over decades to move toward a more environmentally and economically efficient and eventually, if all is done right, sustainable world.

13.3.4   Linking Policy with Long-Term National Goals

For these reasons, it is difficult at best to explicitly establish policies aimed directly at sustainability, which at this point is a long term and ill-defined goal. To the extent they might be effective, attempts to jump directly to sustainability will be fought; to the extent they are ineffectual, they are likely to be ridiculed. Rather, policies should be formulated to achieve shorter term subgoals, and designed so that they do not conflict with, and hopefully contribute to, rather than directly implement, the long term goal. Moreover, policies that are targeted at short term issues should be designed so as to provide incentives for institutional evolution appropriate to the longer term. For example, the U.S. Community Right-to-Know legislation provided that firms had to report data on their emissions to the Environmental Protection Agency, which would make that information public, the short term intention being to inform the public about potential exposure. The desirable institutional change in behavior, however, occurred when firms, facing significant publicity about such emissions, began internalizing pollution prevention and waste reduction as desirable dimensions of production.

Such institutional change tends to occur slowly, but is much more effective in the longer term.

13.4   RELEVANT DIMENSIONS OF NATION STATES

In the case of industrial ecology policy, it is possible to identify, at least at a high level, a number of dimensions along which nation-states fall that can significantly impact their ability to respond to environmental challenges. A few of these, which have proven to be relevant, are worth emphasizing.

13.4.1   Form of Government

In general, it is apparent that democracies, such as those in Western Europe and the United States, have proven to be more responsive than more totalitarian governments, such as those that used to exist in Eastern Europe or the former Soviet Union (FSU). This is not entirely unexpected, because polls have demonstrated that, virtually around the world, environmental issues are a significant concern to a majority of people. Where they can express their opinions through the political process, therefore, one would anticipate a government sensitive to the environment, to the extent its resources, economic, and cultural conditions permit. Another factor, which might be important here, is that authoritarian systems frequently permit ruling elites to expropriate resources, including resources that can be generated from processes causing environmental degradation (e.g., the military rulers in Nigeria grow rich from oil revenues, despite the environmental problems petroleum production has caused in that state).

13.4.2   Wealth

It is axiomatic that developed countries have more resources with which to respond to environmental challenges than poorer countries, and may be able to invest more in enhanced environmental conditions as an element of their higher quality of life than less well off developing countries. Moreover, the process of development, which usually uses existing technological systems, is sometimes achieved by devesting in environmental protection and investing in industrial production. Note, however, that this relationship between development and environmental progress, well illustrated in Taiwan, China, Thailand, and elsewhere, assumes the old paradigm that environmental protection is only a drag on increased productivity, which, in fact, end-of-pipe technologies are in most cases. To the extent that industrial ecology principles are implemented, and environmentally and economically more efficient practices result, development need not imply environmental degradation; in fact, it may well imply the opposite. While this principle is increasingly recognized in places such as China, the will and ability of the state to implement it is less clear. In particular, to the extent implementing it requires that the state be able to pressure investing firms to utilize global best practices (e.g., economically and environmentally efficient processes, product design, packaging, logistics systems, and so on), it may be, in the short term, an option only for those states where the nascent market opportunities are sufficient to justify the firm acceding to such pressure.

There are two aspects of environmental management that are manifestly more difficult in a poor state. The first is the provision of basic environmental management infrastructure: clean water services, sewage systems, and so forth. Privatization of these functions, frequently involving foreign-based transnational corporations, is, in many cases, both feasible and desirable (the foreign direct investment includes experienced management), but may be politically difficult, both for cultural and economic reasons (e.g., the public systems are grossly overstaffed, and privatization implies more efficient operation, including reductions in numbers of employees). The second, of course, involves those very poor countries, and subgroups within countries, characterized by subsistence economic behavior. They may have no choice but to have an impact on environmentally sensitive areas if they wish to live.

13.4.3   Size of Market

The market power of a state is a critical determinant of the structure of its environmental policies. Even very progressive small countries, such as Sweden or the Netherlands, cannot overlook the fact that much of their industrial production is exported, and thus subject to standards and requirements beyond their direct reach. Moreover, if they were to unilaterally change their standards, few transnationals would be impressed; given global markets, losing a small market segment would in most cases have little effect. Compare, however, the ability of Germany, a large producer and consumer, to encourage significant changes in global product design and management in the automotive and electronics sectors through domestic policies such as post-consumer product takeback. The foreign policies of each type of nation accordingly will be different: The Netherlands, for example, has become very good at encouraging the implementation of environmental policies it supports through multinational fora such as the United Nations Environment Programme, the European Union, and the Organization for Economic Co-operation and Development.

13.4.4   Issue

Different cultures and countries emphasize different aspects of environmental protection in their programs. For example, the United States is a leader in remediation, but lags far behind Japan in energy efficiency, Germany in developing consumer takeback approaches to encourage environmentally appropriate design practices, and the Netherlands in environmentally efficient industrial policy development and implementation. There is also the obvious difference in environmental pressures between developed states, where clean air and water, and adequate sewage systems tend to be the norm, and developing states, which generally must focus their efforts on fundamental environmental infrastructure. Figure 13.2 illustrates some of these differences, and their relationship to the wealth of the society.

13.4.5   Culture and Ideology

It should be apparent to even the most superficial observer that there is a significant inherent difference in the way different cultures approach environmental issues. There

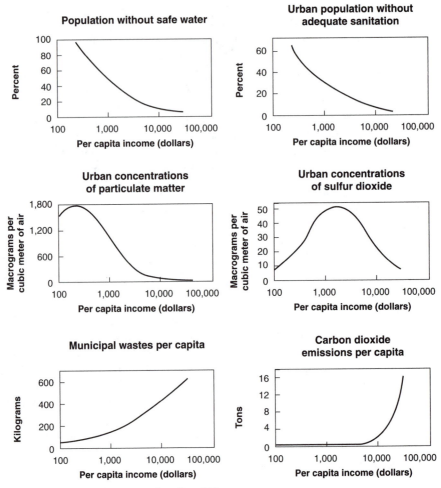

FIGURE 13.2   State of development and quality of life

is a distinct contrast between Japan, with its parsimonious approach to resources, especially energy, born out of its island status, and the United States, with its far greater resource base and accompanying cavalier attitude towards conservation. Bob Ayers, in exploring these issues, refers to the "cowboy economy," a term coined by Kenneth Boulding, in which the government is viewed as an ally in the taming and exploitation of wilderness, an approach that strongly resonates with many in countries such as the United States, Canada, and Australia, but is virtually incomprehensible in countries such as the Netherlands, Sweden, and Germany.

13.4.6   Relationship Between Private and Public Sectors

Another important differentiator is the relationship between the private and public sectors. Both the vision and the practice differ widely among countries that superficially adhere to the capitalistic, free market approach. Japan, for example, has a relatively small government sector, but it is highly interventionist: Japanese ministries conduct their policies in conjunction with, and through, the private sector. France has a relatively large governmental sector but is also highly interventionist: major French firms in many cases are controlled, directly or indirectly, by the state, and the same people shift between high posts in both institutions. In both countries, private and public interests are seen as much more integrated than in a country like the United States, which tends to regard the public and private interests as not just different, but adversarial: government is a necessary counterweight to self-interested private firms. Obviously, the cultural model of the private firm is also different; unlike the case in the United States, private firms in Japan or northern Europe are viewed as having some social responsibilities.

13.4.7 Factors Affecting Development

Not only the current economic status of countries, but in many cases their ability to develop economically is an important variable affecting their approach to environmental and sustainability issues. Recent studies by the Harvard Institute for International Development have indicated that global patterns of growth over the past two decades appear to have depended on four factors: baseline economic status, physical geography, government policies, and demographic change. The role of demographics and baseline economic status are obvious, but some of the other factors are less recognized. Thus, for example, landlocked countries grew, on average, 0.7 percent more slowly than coastal ones, and tropical countries grew 1.3 percent more slowly than those in temperate zones (Jeffrey Sachs of Harvard suggests that this may reflect costs associated with poor health and unproductive farming practices). Government policies are increasingly recognized as a critical contributor to economic development: Developing countries with open economies (e.g., many in Asia) grew 1.2 percent per year faster than those with closed economies (many in Africa and, until recently, in Latin America). Prudent economic policies and plain "good government"—the rule of law, not arbitrary and corrupt elites—are a critical part of capacity building for developing countries (and their lack explains the sudden recent collapse of several Asian economies as well).

Such findings matter because, as seen in many guises throughout this text, the shift of environment from overhead to strategic means that, in many cases, the most effective policies and methodologies in terms of environmental efficiency have little to do with traditional environmental policy. In this case, for example, it may be that the most sophisticated and powerful approach for achieving environmental improvements is the implementation of policies that build capacity for economic development in poorer areas of Asia, Latin America, and Africa. The World Bank and foreign direct investment (FDI), properly informed, may be the most powerful engines for fundamental environmental improvement that we have.

13.5 REGULATORY STRUCTURE

Although each jurisdiction will generate its own regulatory structure to meet its requirements, it is possible to outline a generic policy structure. One element of such a structure consists of developing environmental components of all other appropriate policy structures, such as trade policy, national security policy, and economic policy.

TABLE 13.1 Generic Environmental Management Structure.
Both traditional command-and-control, and boundary setting regulatory approaches have their place, but many jurisdictions and stakeholders tend to rely almost completely on the former.

Regulatory Structure	Policy Drivers	Examples of Policy Tools
Level 2: Setting boundary conditions	Generic improvement of environmental efficiency	Fiscal incentives Reporting and informations systems
Level 1: Command-and-control	Integration of environment with other social goals Significant and difficult to reverse effects Socially unacceptable behavior	Material bans Prescribed control technologies

This simply reflects the transition of environmental issues from overhead to strategic for society, which implies that environmental considerations become integrated into other policy systems that, heretofore, had not needed to consider them.

The second element, of course, reflects a change in the environmental regulatory system itself, which generally consists of two components (Table 13.1). The first is a baseline command-and-control component that responds to two kinds of behavior:

1. That which raises the possibility of such fundamental and irreversible environmental damage, at such a geographic and temporal scale, that it cannot be permitted (e.g., releasing CFCs, which degrade the stratospheric ozone layer);
2. That which is beyond the moral boundary of socially acceptable behavior (e.g., releasing heavy metals from manufacturing cleaning operations into drinking water sources).

Specific examples of such regulatory approaches would be bans on specific materials or activities, such as lead in gasoline, or mandating certain kinds of control technologies, a common regulatory approach under clean air and clean water statutes in many countries.

But as a generic policy structure, the command-and-control approach has significant limitations that argue for sparing use. Command-and-control regulations almost always have unintended consequences, because they presuppose a simple and understood system (see Chapter 9). Even the obvious success stories have some costs: banning lead in gasoline, for example, led to less efficient automobile engines in the short term, because engine compression had to be reduced. Many of the aqueous cleaning systems that the electronics industry substituted for those based on CFCs required more energy per unit product.

Moreover, every command-and-control tool reduces the flexibility of technology systems. Imposing specific emission control technologies, for example, not only freezes innovation in control technology, but may keep economically and environmentally inefficient technologies in use longer. Rigid definitions of "waste," accompanied by regulatory burdens, increasingly pose a major impediment to environmentally preferable materials management regimes. Any regulatory tool that limits the ability of technological systems to adapt and evolve, which the study of industrial ecology indicates is fundamental to the discontinuous improvement in environmental and economic efficiency required for sustainability, must be used sparingly. In this regard, proposals to ban materials, especially widely used materials such as chlorine, without rigorous and comprehensive analysis of alternatives, must be regarded as highly problematic.

Some, especially in industry, use the undeniable inefficiency of command-and-control regulation in many instances to argue that, given appropriate incentives, industry should be freed from direct regulatory requirement. This is an untenable argument. First, it presupposes that economic incentives for environmentally inappropriate behavior have been identified and reduced below the incentives for positive behavior, which is not the case. Second, even if it were, environmentally unacceptable practices are embedded in current technology systems, and would not be dislodged simply by changed incentive structures. Third, such a system is highly unlikely to be accepted by the public in the foreseeable future: industry simply isn't trusted, especially when it comes to environmental performance.

There is, accordingly, the need to better understand and develop the second policy, setting boundary conditions within which private behavior, including market behavior, occurs. This is more difficult, in part because policymakers have little experience with such tools. The exception is the suite of economic tools, including subsidies, taxes, fees, pollution charges, and others, all of which have the goal of internalizing externalities. Although fairly well understood in theory, these suffer from two principle drawbacks: they are difficult to implement politically, and quantifying the externalities is both theoretically and practically challenging.

A more interesting set of issues is raised by tools, such as product eco-labels or emissions reporting for facilities, that have the effect of providing more information to the market, either about the specific product or the generic performance of the firm. These have the potential to be quite powerful as well, and facilitate, rather than substitute for, the operation of the market system. This makes them both efficient and desirable to those who appreciate the efficiency of markets. The principle problem with this suite of tools is that they are conceptually difficult. In most cases, there are not enough data, and existing methodologies are too primitive, to ascertain what product attributes are preferable, and especially to do so with metrics that are consolidated enough for public understanding. Thus, industry professionals tend to be somewhat distrustful of comprehensive product eco-labeling schemes, even while appreciating their potential.

An exception to this concern is single characteristic eco-labels, especially where the dimension being evaluated is sufficiently separate from other product characteristics that there can be some hope that improvements in it are, in fact, equivalent to improvements in product environmental performance. An example might be the U.S. Department of Energy and Environmental Protection Agency's Energy Star labeling program for personal computers, which identifies energy efficient machines. Energy consumption is both easily measured, and relatively independent of other product characteristics. Accordingly, a number of other countries are adopting the Energy Star standards, ensuring both global pressure for better product performance, and harmonization of standards. The latter, in a global market, can contribute substantially to efficient product design and operations.

Another information tool, technical assistance, targets not consumers, but producers. This is a valuable concept, because, in practice, there are usually a number of technologies existing at any given time that are both environmentally and economically more efficient than existing practices. Lack of information, especially among small businesses, tends to inhibit the diffusion of such technologies; this can be overcome by targeted programs. A practical problem with some of these programs is that they are run by environmental regulators, which has two drawbacks. First, many small businesses have a profound distrust for environmental regulators, who in most cases have the institutional responsibility for both enforcement and compliance. Second, environmental regulators are, by and large, not technologists; they tend to be environmental or social scientists, lawyers, or economists. Accordingly, they are not regarded as reliable sources for advice on manufacturing technologies or product design.

A third set of boundary conditions may be thought of as changing the rules under which firms operate. Thus, for example, imposition of post-consumer product takeback requirements imposes a fundamental change on firms. They no longer evaluate their role as simply producing the best product they can as inexpensively as they can, but they must now reduce end-of-life management costs as well, including the logistics system needed for reverse product distribution, the most efficient dismantling technology, and management of any disposal operations to minimize liability, for which they are now responsible. More subtly, takeback systems push firms toward the functionality economy, where consumers purchase function from service firms, rather than products from manufacturing firms. Leasing becomes more desirable as the mechanism by which firms interact with their customers under such circumstances. Note that this changes many things for a manufacturing firm: its product cost structure, its business plan, its product management system, its sales system, its legal obligations, and the source of its revenue stream. It no longer prospers just by making and selling a lot of widgets; rather, it prospers by how well it has designed its widgets across the life cycle to support a service offering. It is these kinds of deep and profound shifts in economic behavior and corporate culture that properly implemented changes in boundary conditions can generate.

The U.S. Congress Office of Technology Assessment reviewed existing environmental policy instruments for Congress, concluding that tools with fixed targets, which would be in level one of Table 13.1, generally are effective at meeting goals, but are not adaptable, and may not be cost effective (Figure 13.3). This appears to hold, however, only when the goal is defined in terms of a simple system; such tools are generally ineffective and inefficient ways to manage toward more complex goals such as an environmentally and economically more efficient economy. Moreover, the process behind policy instruments is also a critical consideration that is, perhaps, not adequately recognized in countries, such as the United States, with adversarial regulatory systems. The rigidity imposed by "notice and comment" rulemaking, required in many U.S. environmental programs, contrasts significantly with the greater efficiency of the covenant system used in the Netherlands.

In reality, as Bill Long of the OECD (Organization for Economic Co-operation and Development) points out, policy tools will fall along a continuum (Figure 13.4). Which one is appropriate will depend on the concern that the policy is intended to

Policy Instruments

Legend:
- ■ Effective
- ● It depends
- △ Use with caution
- ★ Average

	Environmental results			Costs and burdens		Change	
	Assurance of meeting goals	Pollution prevention	Environmental equity and justice	Cost-effectiveness and fairness	Demands on government	Adaptability	Technology innovation and diffusion
Tools *without* fixed targets							
Technical assistance	△	■	■	●	★	■	★
Subsidies	△	★	■	★	△	★	★
Information reporting	△	★	■	●	■	■	★
Liability	★	●	★	★	★	■	★
Pollution charges	★	★	△	★	★	★	■
Tools *with* fixed targets-multisource							
Challenge regulations	★	★	△	●	●	●	●
Tradeable emissions	●	★	△	■	★	★	●
Integrated permitting	■	★	★	●	★	★	★
Tools *with* fixed targets-single-source							
Harm-based standards	■	★	★	★	△	★	★
Design standards	■	●	★	★	★	△	★
Technology specifications	■	●	★	△	★	△	★
Product bans	■	■	★	△	★	△	■

FIGURE 13.3   Policy Instruments. Tools with fixed targets-single-source tend to be used most frequently. While they offer a degree of certainty, they are ill-adapted to complex systems where tools without fixed targets are most effective. To assure public acceptance, the latter need to be supported by methods for externally validating performance against objective metrics.

Source: U.S. Office of Technology Assessment, *Environmental Policy Tools*, 1995

address, and the internal and external dynamics that the jurisdiction faces. It is also apparent that, taken across developed countries as a whole, policy sophistication is improving over time (Figure 13.5), although individual national states may be more or less advanced.

As in the case of sector and technology initiatives such as life-cycle assessment (LCA) and Design for Environment (DFE), it is a period of experimentation and trial for policy options as well. Moreover, the shift away from treating environment simply as a social overhead issue, and the increasing recognition of the critical role of economics in achieving improvements in environmental performance, have begun to generate an impressive, if confusing, tapestry of policy experimentation around the world. The World Bank Environmental Department in its publication *Five Years After Rio: Innovations in Environmental Policy* has captured some of this—it can only be called exuberance—in a policy matrix, reproduced as Figures 13.6 through 13.9. These

Continuum of Degrees and Types of Government Intervention

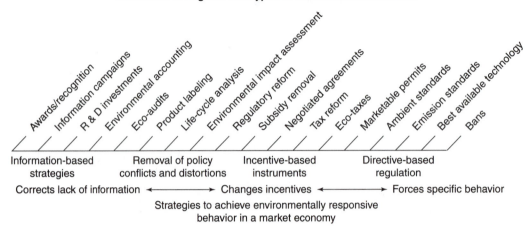

FIGURE 13.4 Continuum of degrees and types of government intervention. Regulations utilizing mechanisms on the left and middle of the scale are more appropriate for managing complex systems than those on the right.
Source: Bill Long, OECD, presented at ECO '97 International Congress, Feb. 24–26, 1997, Paris

Evolution of Environmental Policy in OECD Countries

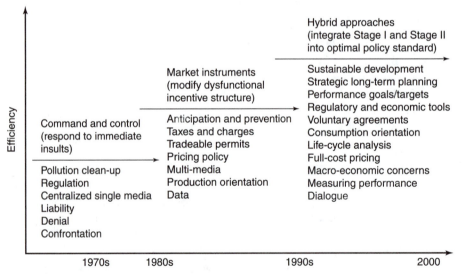

FIGURE 13.5 Evolution of environmental policy in OECD countries. In practice, many countries retain a regulatory system centered around 1970's mechanisms.
Source: Based on Bill Long, OECD, Presentation at ECO '97 International Congress, Feb. 24–26, 1997, Paris.

efforts, which are matched at local jurisdictional levels and complemented by similar activities on the part of transnationals and NGOs, are laying the necessary experiential groundwork for the policy structures that will be required if sustainability is to be more than a slogan.

The Policy Matrix—Instruments and Selected Example Applications

	By sector or theme	Using markets: Subsidy reduction	Environmental taxes on: emissions	Environmental taxes on: inputs	Environmental taxes on: products	User fees for: natural resources	User fees for: services	Perform. bonds/ deposit-refund	Targeted subsidies
Resource management	Water resources	reduction in water subsidy: Hungary, Poland, China				water resources taxes: Germany, Brazil	• water pricing: Chile, Colombia, China		
Resource management	Fisheries			fishing input taxes	product taxes	fisheries licenses: Mauritania		oil spill bonds: US	
Resource management	Land management	Removal of land conversion subsidies: Brazil Removal of tax on undeveloped land: France				• property taxes • differential land use taxes/fees: Germany	betterment charges: Korea, Mexico		subsidies for restoring natural cover: Canada
Resource management	Forests	reduction in land conversion subsidies and subsidized livestock credit: Central America, Brazil			forest product taxes: Brazil, Colombia, Venezuela	stumpage fees: The Philippines, Indonesia, Malaysia, Brazil, Costa Rica, Honduras	• park entrance fees: Costa Rica • reforestation taxes: Indonesia	reforestation deposit/ performance bond: Indonesia, Malaysia, Costa Rica	• subsidies for seedlings: India • credit subsidies for reforestation: Costa Rica
Resource management	Sustainable agriculture	reduction in agri. subsidies: most developing countries							
Resource management	Biodiversity/ protected area	reduction in land conversion subsidies				bioprospecting fees: Costa Rica, Madagascar	• watershed protection charges: Indonesia, Costa Rica • park entrance fees: Indonesia, Costa Rica, Nepal		habitat protection subsidies
Resource management	Mineral resource		fees on mine wastes and tailings: Philippines			mining royalities: Canada, U.S., other OECD, Algeria, Brunei, Brazil, Columbia, Ecuador, Venezuela, Malaysia, Namibia, Nigeria			

FIGURE 13.6 Policy matrix

The Policy Matrix—Instruments and Selected Example Applications

Pollution control

By sector or theme	Subsidy reduction	Using markets — Environmental taxes on: emmissions	Environmental taxes on: inputs	Environmental taxes on: products	User fees for: natural resources	User fees for: services	Perform. bonds/ deposit-refund	Targeted subsidies
Air pollution	reduction in energy subsidies: transition economics, most developing countries	emission taxes: OECD, Egypt, Korea, China, Eastern Europe, Kazakstan	• energy taxes: OECD countries • differentiated gasoline prices: Philippines, Egypt, Turkey, Mexico	environment-related product taxes: OECD, Bangladesh	royalties for fossil fuel extraction		sulfur tax refund systems: Sweden	subsidies for industrial energy-saving: Sweden
Water pollution		wastewater discharge fees: OECD, China, Brazil, Mexico, Korea, Philippines, Eastern Europe				sewage charges: Indonesia, China, Malaysia, Thailand, Brazil, Colombia, Mexico, Singapore, Chile		tax relief and subsidized credit for env. investment: India, Korea, China, Philippines, Brazil, Mexico, Colombia, Chile, Ecuador
Solid waste		waste disposal taxes: France, UK, Canada, U.S.		product taxes: Denmark		user fees for waste management: OECD, Ecuador, Venezuela, Thailand	deposit-refund systems: U.S., Finland, Norway, Sweden, Japan, Taiwan (China), Philippines, Bangladesh, Brazil, Chile, Colombia, Ecuador, Jamaica, Mexico, Venezuela	credit/subsidy policy: Korea, Taiwan (China), Turkey, UK
Hazardous waste/toxic chemicals	reduction in agro-chemical subsidies: most developing countries	disposal charges: OECD, China, Thailand	pesticide taxes: OECD countries				bonds for waste treatment	subsidies for phasing-out pesticides: Sweden

FIGURE 13.7   Policy matrix

The Policy Matrix—Instruments and Selected Example Applications

By sector or theme	Creating markets			Environmental regulations			Engaging the public	
	Property rights/ decentralization	Tradeable permits/ rights	International offset systems	Standards	Bans	Quotas	Information disclosure	Public participation
Water resources	• water rights: Chile, U.S. • decentralization of water supply management: Hungary	water markets: Australia, India, New Zealand, U.S., Chile	water trading across borders	water quality standards	seasonal ban of certain types of water use: U.S.	water use quotas: Israel, U.S.	water efficiency labeling	water user association: Mexico, Argentina, Turkey
Fisheries	200 mile EEZ	tradeable quotas/ permits: New Zealand		fishing standards	ban of fishing	fishing quotas		
Land management	land title: Thailand	transferable development rights: Puerto Rico, U.S.	tradeable conservation credits	land use standards/zoning: OECD, Guatamala, Korea, China, Pakistan, Brazil	establishment of enviro. sensitive areas	land subdivision		private participation: Thailand, Turkey
Forests	land titling	tradeable reforestation credit	tradeable reforestation credits: Costa Rica, Panama, Russia	• logging regulations • zoning: Brazil	logging/log export bans: Costa Rica	logging quotas: Malaysia	ecolabeling: Nordic countries	
Sustainable agriculture	• land ownership: Thailand • participatory irrigation management: Mexico, Argentina, Philippines, Sri Lanka, Tunisia	transferable development rights		zoning: Brazil	ban on use of pesticides: Indonesia, Latin America		ecolabeling: many OECD countries	community self-help groups: Kenya
Biodiversity/ protected area	biodiversity patents and bioprospecting rights: Costa Rica, Madagascar	international tradeable conservation credits: Costa Rica, Mexico	tradeable conservation credits	conservation zoning: Brazil, Costa Rica, China	establishment of national parks: Indonesia, Brazil, Costa Rica			NGO involvement: Philippines
Mineral resource				waste and tailings containment	ban of mining		energy efficiency labeling: Australia	

Resource management

FIGURE 13.8   Policy matrix

The Policy Matrix—Instruments and Selected Example Applications

<table>
<thead>
<tr>
<th rowspan="2">By sector or theme</th>
<th colspan="3">Creating markets</th>
<th colspan="3">Environmental regulations</th>
<th colspan="2">Engaging the public</th>
</tr>
<tr>
<th>Property rights/ decentralization</th>
<th>Tradeable permits/ rights</th>
<th>International offset systems</th>
<th>Standards</th>
<th>Bans</th>
<th>Quotas</th>
<th>Information disclosure</th>
<th>Public participation</th>
</tr>
</thead>
<tbody>
<tr>
<td>Air pollution</td>
<td>• environmental liability
• private energy production: Philippines</td>
<td>• tradeable emission permits: U.S., Chile, Poland, Czech Republic, Kazakstan
• auctionable permits for ozone-depleting substances: Singapore, Mexico</td>
<td>Joint implementation carbon offsets: U.S., Argentina, Norway, Latin America, Russia, Poland</td>
<td>air quality and emission standards: OECD, India, Brazil, China, Turkey, Korea, Singapore, Philippines</td>
<td>ban on imports of ozone depleting substances: China</td>
<td>• emission quotas: OECD
• area licensing for vehicles: Singapore</td>
<td>public disclosure program: Indonesia, Chile</td>
<td></td>
</tr>
<tr>
<td>Water pollution</td>
<td>environmental liability</td>
<td>tradeable wastewater discharge permits</td>
<td></td>
<td>wastewater discharge standards: OECD, India, Indonesia, Malaysia, China, Korea, Singapore, Philippines</td>
<td></td>
<td>industrial wastewater discharge quotas: OECD, Colombia, Bahamas, China</td>
<td>public disclosure program: Indonesia, Bangladesh, Philippines</td>
<td>community pressure: Korea</td>
</tr>
<tr>
<td>Solid waste</td>
<td>environmental liability</td>
<td></td>
<td>tradeable wastewater discharge permits</td>
<td>landfill standards and landfill zoning</td>
<td></td>
<td></td>
<td>industrial waste exchange program: Philippines</td>
<td></td>
</tr>
<tr>
<td>Hazardous waste/toxic chemicals</td>
<td>environmental liability</td>
<td></td>
<td></td>
<td>containment/ treatment standards</td>
<td>• bans of international transportation: Basel Convention
• ban on use of some pesticides: Indonesia</td>
<td></td>
<td>labeling</td>
<td></td>
</tr>
</tbody>
</table>

Pollution control

FIGURE 13.9 Policy matrix

REFERENCES

Ayers, R.U. "Cowboys, cornucopians, and long-run sustainability." *Ecological Economics* 8:189–207 (1993).

Bormann, B.T., M.H. Brookes, E.D. Ford, A.R. Keister, C.D. Oliver, and J. F. Weigant. *A Broad, Strategic Framework for Sustainable-Ecosystem Management.* Prepared for the USDA Forest Service, May, 1993.

Davies, J.C. and J. Mazurek. *Regulating Pollution: Does the U.S. System Work?.* Washington, DC, Resources for the Future: 1997.

European Commission. "Communication from the Commission to the Council and the European Parliament on Environmental Agreements." Com(96) 561, Brussels, November 27, 1996.

Sachs, J. "The limits of convergence: nature, nurture and growth." *The Economist.* June 14, 1997, pp. 19–21.

Schmidheiny, S. and the Business Council for Sustainable Development. *Changing Course.* Cambridge, MA, The MIT Press: 1992.

Smart, B., ed. *Beyond Compliance: A New Industry View of the Environment.* Washington, DC, World Resources Institute: 1992.

U.S. Congress, Office of Technology Assessment. *Environmental Policy Tools: A User's Guide.* OTA-ENV-634 Washington, DC, U.S. Government Printing Office, September 1995.

World Bank. *Five Years After Rio: Innovations in Environmental Policy* (Discussion Draft). March, 1997.

Weinberg, M., G. Eyring, J. Raguso, and D. Jensen. "Industrial ecology: the role of government," in B.R. Allenby and D.J. Richards, eds. *The Greening of Industrial Ecosystems* Washington, DC, National Academy Press: 1995, pp. 123–136.

EXERCISES

1. You are the Technology Minister in a developing country that has a number of metal platers in urban areas, most located near rivers heavily utilized for drinking, cooking, and washing. The metal plating sector is an important source of economic benefit for your country, and a significant recipient of foreign direct investment funds. Data indicate, however, substantial and increasing metal contamination of surface waters in urban areas. Facing this situation, what are your policy goals? Prioritize them, and outline and defend a proposed set of policies to achieve them.

2. You have just been appointed the Environmental Minister when your Prime Minister calls you in and tells you that she doesn't like cars with internal combustion engines, and wants fuel cell/flywheel hybrid vehicles instead. What policies would you recommend if you are in the government of:

   a.   the United States?

   b.   Sweden?

   c.   The People's Republic of China?

   d.   Nigeria?

3. You are a policy analyst with an environmental non-governmental organization (NGO) in a developed country. You have been asked to craft a set of policy recommendations that will "make us a sustainable country."

   a.   Write a three page paper laying out and defending your policy recommendations.

   b.   How do your recommendations change if you are in a rapidly developing Asian country?

   c.   How do your recommendations change if you are in a poor African country?

CHAPTER 14

The Private Firm: Managing Industrial Ecology Implementation

14.1 OVERVIEW

The entity known as the private firm is somewhat like the weather: pervasive, critical, little understood— and chaotic. Although the private firm is a creature of law (and thus defined therein), scholars have advanced a number of potential functional definitions from differing perspectives: historical, legal, economic, sociological, management studies, and psychological. More prosaically, to a large extent a firm is an accounting entity. Even within a functional definition, there may be differences: An entity may look like a firm for purposes of security law, but simply be a group of people for tax law purposes. In the case study in Chapter 16 of this text, for example, the firm is treated as a subsystem within a larger complex system, the global economy, that has resulted from the Industrial Revolution. Each of these approaches has its value, and usually catches some of the reality, but they all remain somewhat incomplete.

Even when firms introspect, there is some ambiguity. For example, the author is familiar with a large diversified firm with a number of business units that asked several of its employees to identify the meaningful differences of being inside or outside the firm (i.e., would it make any difference if the business units were all spun off rather than integrated into a single firm). Except for a possible increase in the cost of capital because of potentially higher risk associated with smaller firms, it proved difficult to identify any real difference between a business unit within the firm, and one on its own as an independent firm.

This text is not the place to attempt to resolve what at times becomes an almost theological discussion. Rather, this chapter is intended to encourage the student of policy to think more carefully and with greater sophistication about what a firm really is, with the goal of more environmentally and economically efficient policies as a result. After all, private firms both dominate the development and deployment of technology, and are critical to global economic activity. They are equally critical for progress toward a more sustainable economy. Some knowledge of the relevant dimensions of their day-to-day activity, and the way in which they view and manage change, is therefore quite useful in policy formulation.

This chapter will deal with mundane aspects of corporate activities. The more profound question of whether private firms as currently constituted can support the development and maintenance of a sustainable economy is raised in Chapter 16. While the issues raised here may strike some as simplistic, a realistic assessment of current policies indicates that they are apparently little understood, at least in policy development and implementation.

14.2   THE FIRM AS EVIL BLACK BOX

Many people, including many environmental regulators and policymakers, tend to have mental models of private firms, to a greater or lesser degree, as both monolithic and deliberately evil. These views, while perhaps understandable, are both oversimplistic and inefficient if embedded in regulation. For these reasons, rational policymaking calls for a less ideological, more practical concept of the firm, and an increased understanding of its internal dynamics as it may affect environmental performance.

Like all large institutions, firms are more complex, multidimensional, and even internally contradictory than most people realize. Factions, politics, cultural and operational differences, internal conflict, dysfunctional behavior—all are endemic to the firm. Creating change, such as implementing industrial ecology through such methodologies as Design for Environment, in such circumstances is essentially organizational guerrilla warfare. For those interested in more effective and efficient policy, this is a critical understanding. Policy should seek to identify and support those factions within the firm fighting to achieve the desired changes, and to reduce legal, cultural, and psychological barriers to such changes. This requires considerably more sophistication than the current approach, which essentially considers the firm monolithic in both internal structure and attitude, and essentially treats organizational issues as irrelevant. As Richard Nelson of Columbia University points out, this view of the firm is rooted in standard neoclassic economic theory, which operates at the level of markets, and thus tends to ignore institutional idiosyncrasies and dynamics. Where one is satisfied with a high-level, static analysis, this may suffice; where, as in industrial ecology policy, the goal is to actually encourage evolution of appropriate corporate behavior, it is inadequate.

Two examples should make this point a little clearer. One is the implementation of Community Right-to-Know requirements in the United States as part of the Superfund Amendments and Reauthorization Act of 1986 (SARA), which required facilities emitting over a threshhold amount of designated substances to report these data to the government. At the time, virtually all actors involved in passing SARA focused on other aspects of the legislation. Yet, when the emissions data were submitted and made public by EPA, a significant change in corporate behavior occurred. Plant managers, who had previously tended to regard their facilities as closed baronies, faced significant incentives to change their approach as a result of the publication of emissions data, frequently on the front pages of their local newspapers. The result was both short term—reductions in emissions—and, more subtly, long term—a shift in the way management regarded their facility and its relationship with the surrounding community. Although unintended and unforeseen at the time, the effect of this provision has been to change corporate behavior among a specific, but critical, subset of company management: facility managers and their immediate superiors. It thus remains, perhaps, one of the most effective environmental requirements passed, even though it does not include any enforcement provisions targeted to emissions whatsoever.

A second example is the tax imposed by the United States on CFCs, which went beyond the phaseout required by the Montreal Protocol. The familiar claim for this approach is that, by adjusting economic prices toward social costs, such a tax encourages more efficient behavior. This is true: A phaseout that is several years away, even if Draconian, is less likely to change management behavior than a tax, which predictably increases the cost of a certain course of action. There was also, however, a more subtle institutional effect of this tax, which provides an example of how policy can modify corporate behavior in an appropriate manner. The tax provided a key argument for those personnel internal to manufacturing firms who were urging the development and deployment of non-CFC alternatives and, more broadly, implementation of Design for Environment practices. It gave their arguments critical economic credibility within the firm by providing a business case for urging quantum improvement in integration of environmental considerations into product design, manufacture, and management activities, a dynamic all but invisible to the external world.

Viewing firms—indeed, capitalism itself—as in some sense morally suspect has a long history; one need only recall Marx or the history of the organized labor movement. For environmentalists, this tendency appears to arise from two principle sources: a deep distrust of technology, based on the environmental impacts that the current technological civilization has unquestionably had, and a suspicion of the profit motive and the concomitant perceived amorality of private firms. These are not unreasonable lessons to draw from the history of the Industrial Revolution. They are counterproductive, however, if, rather than encouraging an appropriate caution and skepticism, they institutionalize opposition to all technological evolution and an adversarial approach to private firms and their role in achieving enhanced economic and environmental efficiency.

It is perhaps more productive from a policy development and deployment perspective to adopt the model of "firm as protoplasm": firms adapt to their environment given their internal and external resources and state, and degrees of freedom, just like a blob of protoplasm. The challenge to policy, therefore, is to encourage them to evolve in the right direction by providing appropriate incentives and boundary conditions, while using command-and-control where required to control unacceptable behavior. If this is to be accomplished, the oversimplistic preconceptions that have dominated the relationship between regulators and private firms should be replaced with a more sophisticated view of the firm as a social construct, replete with internal contradictions, which can be manipulated by understanding, and taking advantage of, its complex internal structure.

A logical extension of this more sophisticated view of the firm is the need to beware of those who speak of "industry" as a whole. There is no such animal: it is usually the case that sectors, and firms within sectors, will have very different positions. For example, the extractive and natural resource sectors—mining and petroleum—tend to be much less interested in environmental values (based on their lobbying positions) than more progressive sectors such as electronics manufacturing, automotive, and telecommunications. Moreover, with complex environmental perturbations, it is likely that sectors will be differentially affected, which will also contribute to different perspectives and positions in various industrial segments.

Moreover, within any sector, some firms will be leaders, while some will, at best, be followers. This is true even though there is a "herd mentality" among firms that tends to discourage any single firm from getting too far ahead of others on specific issues. After all, firms are linked in complex supplier/customer networks, and commercial interests can be used to control dissent. This "herd mentality" is also apparent in trade associations, which, by their nature, tend to adopt positions that reflect the lowest common denominator among their members (as most associations must operate on a consensual basis). In fact, presentations of trade associations are one place where the blanket use of the term is common: "Industry is opposed to this measure . . . ." A cynic would say that the weaker the position, the more frequent is likely to be the claim that "industry" as a whole supports it. Very few issues create this kind of consensus.

In fact, it is probably the case that "sustainability issues" increase, rather than decrease, the distance between leading and trailing firms. This has an important policy implication: Policy should seek ways to encourage these differences, identify and reward leading firms, and use their standards of behavior and performance as goads to the rest. When AT&T, for example, became the first electronics firm to publicly commit to eliminate its use of CFCs in electronics manufacturing, it was severely (albeit privately) criticized by a number of other electronics manufacturing firms on a number of grounds. Rapidly, however, a few firms that had been working on such policies internally committed to them as well, and the rest, recognizing the futility, and cost in brand name perception, if they did not do so as well, soon followed. Wise industrial ecology policy recognizes such "lead firm" opportunities—or, indeed, creates them—and uses them to shift blocks of firms to more preferable practices.

14.3   ENVIRONMENT AS STRATEGIC FOR THE FIRM

Just as environmental issues are becoming strategic, rather than overhead, for society (recall Table 1.1), they are becoming strategic for private firms as well (Table 14.1). In the past, environmental activity had been considered to represent simply correction of bad or careless practice. Remediation, regulation, and compliance activities within the firm generally focused on single media, individual substances, specific sites, or particular emission points (down to individual process emissions). There was little recognition (or incentive to recognize) that the impacts arising from the firm's activities were fundamentally linked with regional and global natural, technological and economic systems, and the pervasive appeal of the modern consumer-oriented economy. The policy goal behind environment as strategic for the firm—the notion of sustainability at some meaningful time scale—requires that recognition, however, and a fundamental integration of environment, economic activity, and technology at the level of the firm.

The principle drivers for this activity also shift in this process. Although well-crafted regulation (of which there is too little) remains important, market demand, which tends to be far more important in the mental models of business people, becomes increasingly

TABLE 14.1   Evolution of Private Firm Environmental Management. While managers in leading firms are beginning to wrestle with questions of sustainability, they are generally as confused as other experts about details, particularly metrics and appropriate tools. Moreover, it would be premature to identify any major firm as having fully internalized the implications of sustainability.

Time Focus	Principle Activity	Environment in Firm	Principle Drivers	Policy Goal
Past	Remediation (local scale of activities)	Overhead; function performed by environmental specialist.	Liability; government environmental regulation	Localized cleanup to reduce human risk
Present/Past Focus	Compliance; end-of-pipe controls of specific emissions (local, frequently facility-specific scale of activities)	Overhead; function performed by environmental specialist	Liability; government environmental regulation	Meet substance-specific numerical emission standards
Present/ Future Focus	Design for Environment and other industrial ecology approaches; global scale	Strategic; included in all design, planning and operating organizations and systems	Customer demand; government and world trade product standards and incentives	Sustainability; metrics generally unavailable

powerful. Sophisticated customers that drive much commercial procurement, such as governments in Japan, Europe, and the United States, increasingly demand environmentally preferable products and services, even though defining such offerings remains difficult. ISO 14000, a set of environmental requirements now under development by the International Organization for Standards, is generating a number of organizational methodological and product requirements through a relatively transparent, albeit tedious, negotiating process involving industrial, government, and public stakeholders (Figure 14.1). Eco-labeling schemes, such as the Energy Star for energy efficient personal computers, or the German Blue Angel eco-labeling scheme, impose product design and operation requirements, a far cry from end-of-pipe controls on manufacturing facilities.

The Blue Angel requirements for personal computers, for example, include modular design of computer systems, customer replaceable subassemblies, avoidance of bonding between incompatible materials such as plastics and metals, and post-consumer product takeback. Implementing takeback requirements alone requires development of a reverse logistics system, to get the products back to a central location where they can be disassembled; establishing a new relationship with suppliers, as they become responsible for their piece of the returned product, or at least for appropriate initial design of their components or subassemblies; developing a new corporate capability

ISO 14000 Environmental Standards System

Standards area	Series	Description
Management systems	14001	Describes basic elements of an *environmental managment system* (EMS)
	14004	Guidance document explaining and defining key environmental concepts
Evaluation and auditing	14010	Guidelines on general environmental auditing principles
	14011	Guidelines on audit procedures, including audits of EMSs
	14012	Guidelines on environmental audit or qualifications
	14013	Guidelines on managing internal audits
	14014	Guidelines on initial reviews
	14015	Guidelines on site assessments
	14031	Definition of, and guidance on, environmental performance evaluation
Environmental labeling	14020	General principles
	14021	Terms and definitions for self-declared environmental claims
	14022	Symbols
	14023	Testing and verfication
	14024	Criteria for product evaluation and label awards
Environmental assessment of products	14040	General principles and guidelines
	14041	Guidelines for life cycle assessment impact assessement
Environmental aspects in product standards	14061	Guidance for writers of product standards

FIGURE 14.1   ISO 14000 Environmental Standards System

(disassembly and management of end-of-life products and material streams); and, con-comitantly with these efforts, a completely new concept of product life-cycle cost and business planning process. It is obvious, of course, that none of these requirements have anything to do with traditional environmental approaches, but rather involve the design, business planning, marketing, strategic planning, and technology choice activi-ties that are the core of any manufacturing company. And the overall impact on indus-trial behavior is equally profound, as the shift of industrial firms from manufacturing to service orientation strengthens as a result.

Moreover, failure to effectively perform these functions does not, as with tradi-tional environmental regulation, simply expose the firm to liability, however undesir-able that might be. Rather, failure to perform runs directly to the firm's ability to market its product, and the cost structure of each product. Failure to design logistics systems, business and marketing plans, and products for post-consumer takeback, and efficient disassembly and recycling, does not simply raise a firm's overhead. It prices a firm's products out of the market, regardless of how efficiently the product can be made initially. Environmental capabilities change from being just a way to control lia-bilities to becoming a source of sustainable competitive advantage.

14.4   IMPLEMENTING INDUSTRIAL ECOLOGY

Under these circumstances, it is perhaps understandable that in many cases the kinds of organizational changes required to implement the principles of industrial ecology in existing firms implies fundamental culture change. This is true both within the firm's existing divisions and organizational entities, and externally, as the firm seeks to achieve new relationships with customers and suppliers alike. While it is true that new, more environmentally and economically efficient technologies are an important support for this evolution, it is also true in many firms that the most difficult barriers to implementing such activities are cultural, not technological.

In most cases, therefore, it is necessary to implement all of the activities usually associated with culture change in complex organizations:

- identifying and supporting champions, who ideally come from appropriate organizations within the firm, and are willing to take the associated risks (which, contrary to idealistic doctrine, are usually substantial);
- identifying and eliminating barriers to change, including those that arise from corporate culture and informal patterns;
- finding the least threatening method of introducing new techniques, tools, and systems (the best changes are those that are never recognized by those who implement them);
- developing strong rationales for new activities that are defined in terms of the target audience's interests and culture.

Thus, for example, it would not necessarily be optimal to introduce environmentally appropriate accounting systems through the traditional environmental organization. Potential users would doubt, and rightfully so, that the environmental organization was the appropriate corporate source for accounting expertise. Rather, one would find a champion within the Chief Financial Officer's organization who would introduce such a system, preferably as just a part of another desirable system such as, in this case, Activity Based Costing (ABC).

The evolution of environmental issues from overhead to strategic also completely changes their role within the firm. Rather than dealing with them through a single, overhead organization (as is still the case in many firms), environmental expertise must be diffused throughout the firm, becoming part of the accounting process, the strategic and business planning process, the research and development process, the product design process, and the marketing process.

Moreover, the competencies required to deal with environmental issues shift as well. Traditional environmental personnel are highly specialized with precise and well-defined responsibilities, chosen more for detailed knowledge than for business or strategic aptitude. These skills do not prepare them to play a leading role as environmental issues become strategic for the firm. Under these conditions, it is ironic but not surprising that many firms find that their environmental organizations may be among the most opposed to this shift in treatment of environmental issues.

At the present time, no firm has fully implemented the principles of industrial ecology, be it Design for Environment (DFE) for a manufacturing firm, or Integrated

Pest Management (IPM) for a large corporate agricultural operation. Indeed, given the existing state of knowledge, it is probably impossible for any firm to do so, even theoretically. However, it is useful to understand the steps that some firms are taking in this direction, as policies which support such activities, or at least reduce the internal barriers to them, would be highly desirable.

14.4.1   Establish Tactical Organizational Structures

Although it varies by firm, the traditional environmental organizations will likely be ineffectual if given the complete mission of implementing industrial ecology principles, for the reasons discussed above. An environmental organization that purports to establish product design criteria will have little credibility with product design teams, although their input, properly presented, can help the team generate a preferable design. An environmental organization that issues green accounting standards to the firm will have a difficult time achieving credibility with business and accounting managers, although they can provide the data that support the accounting system. Accordingly, a corporate entity that is not perceived as part of the core environmental group is frequently more effective.

More subtly, this association of organization with capability explains why compliance technology transfer operations of an environmental regulatory agency may succeed, but clean manufacturing technology transfer operations run by the same entity may not. In the one case, business customers will perceive a link between the proffered function and the entity's competence, while they will perceive a lack of linkage in the other. In the U.S., for example, EPA is generally not regarded by industry as a reliable source of information on manufacturing technologies, while the Department of Commerce is.

This also explains in part the need to support development of technological sophistication in environmental NGOs (non-governmental organizations), who are by and large marginalized when dealing with the technological community because they are not perceived to be credible even when their ideas, in fact, are reasonable.

14.4.2   Establish Training Programs

Many firms, especially medium and small size firms, have little internal capability to comply with existing command-and-control environmental regulations, much less develop and use industrial ecology methodologies such as Design for Environment. Thus, there is an important role for establishing training programs that not only present specific tools appropriate to the sector, but also the broader context within which such practices can be shown to make business sense. While large firms will in many cases do this for themselves, there is a clear role for government support in other cases, such as with sectors characterized by many small companies (printing, for example). Again, however, the location of the training function is important. If it is offered by the Environmental Protection Agency, for example, smaller firms will both 1) be concerned about the perceived integration of training and compliance functions, and 2) tend to question the technological expertise of the training. It is better to have an organization perceived as independent and technologically sophisticated—local universities and community colleges, for example—involved in delivering the training.

14.4.3   Establish Technical Support

Integrating environmental considerations into all aspects of a firm's business is difficult enough. It becomes virtually impossible unless the burden on those who are supposed to do so is minimized. This can be done by providing technical support of many kinds: information on environmentally and economically preferable technologies, information on life cycle and DFE methodologies, information on relevant trade or marketing requirements. In large firms, this capability can be developed internally, perhaps in conjunction with the training program. In other cases, however, policies to offer such services through governmental or traditional educational means should be implemented. Thus, for example, if a country has a series of publicly funded technology centers, these can be upgraded to support DFE, LCA, and other methodologies, and to provide information on environmentally and economically more efficient materials, processes, and product design choices. Alternatively, academic institutions can be funded to provide such services. If done properly, this approach has the advantage of training students in this area as well.

14.4.4   Generate Initial Successes

The old saw that nothing succeeds like success is particularly applicable when significant change to the status quo is desired. The factions within firms that support DFE or equivalent practices will frequently try to find projects that can provide an initial, and visible, success. Similarly, policy should attempt to do the same by, for example, choosing problems that are amenable to relatively easy solution and involving firms or sectors that are likely to be cooperative. Thus, for example, USEPA works frequently with electronics firms or the sector's trade associations on DFE-type projects not because they believe that electronics has more impact on the environment than mining or petroleum, but because the sector is both among the most progressive and the most cooperative on industrial ecology issues, which greatly enhances the opportunity for early successes. Another related example is the USEPA Energy Star eco-labeling program, a success in part because it relied on the electronics industry for initial stages of label development, and because it focused on a relatively defined and easily measured parameter of product performance, energy consumption during the use lifecycle stage.

REFERENCES

Allenby, B.R. "Design for Environment: managing for the future," in G.A. West and R.M. Michaud, eds. *Principles of Environmental, Health and Safety Management.* Rockville, MD, Government Institutes: 1995.

Allenby, B.R. and D.J. Richards, eds. *The Greening of Industrial Ecosystems.* Washington, DC, National Academy Press: 1995.

Castells, M. and P. Hall. *Technopoles of the World: The Making of 21st Century Industrial Complexes.* London, Routledge: 1994.

Dertouzos, M.L., R.K. Lester, R.M. Solow, and the M.I.T. Commission on Industrial Productivity. *Made in America: Regaining the Productive Edge.* New York, Harper Perennial: 1989.

Gladwin, T.N., T.K. Freeman, and J.J. Kennelly. *Ending our denial and destruction of nature: towards biophysically sustainable management theory.* New York University, Stern School of Business, May 1994.

Nelson, R. *The Sources of Economic Growth.* Cambridge, MA, Harvard University Press: 1996.

Scherer, F.M. *Industrial Market Structure and Economic Performance.* Boston, Houghton Mifflin Company: 1980.

Schmidheiny, S., and the Business Council for Sustainable Development. *Changing Course.* Cambridge, Mass., The MIT Press: 1992.

Smart, B. *Beyond Compliance: A New Industry View of the Environmen.t* Washington, DC, World Resources Institute: 1992.

Vagts, D. F. *Basic Corporation Law.* Mineola, New York, The Foundation Press, Inc: 1973.

EXERCISES

1. You are the newly assigned DFE guru for a major manufacturer of furniture with an average compliance record and growing concern for the environment. You have been given the assignment of implementing DFE in the firm within a year.

    a. Develop a plan, including actions, milestones, and metrics where appropriate, for completing your assignment.

    b. Does your plan change when you learn that the company expects to be exporting 90 percent of its production to Europe within five years?

2. It is 1986. You are the newly assigned USEPA guru for ozone depletion. Your assignment is to stop American electronics companies from using CFCs, which contribute to this environmental perturbation. You are also asked to contribute to getting the Montreal Protocol finalized; a number of developing countries are indicating that they may not agree to it because they are afraid their local electronics manufacturing operations cannot succeed without CFCs. You also know that this industry is one with several firms that are environmental leaders.

    a. Develop a plan, including actions, milestones, and metrics where appropriate, for completing your assignments.

    b. Assume that your plan relies in part on identifying a leading firm that can be pried away from its compatriots and used as an exemplar. How would you identify such a firm? How would you encourage it to break ranks with its fellows?

3. You are the new Environmental Minister of a country that has just decided to become sustainable in one generation (25 years). You are sophisticated enough to know that you must rely principally on market mechanisms, and the evolution and diffusion of new, environmentally preferable, technologies by private firms. Using your insight into private firms, what non-traditional (i.e., not end-of-pipe command-and-control) policies would you implement? Give several examples of such policies in action, and evaluate whether they have been effective.

P A R T   I I I

CASE STUDIES

Structured Design for Environment Case Study: The AT&T Matrix System

15.1 OVERVIEW

This text has emphasized two related and critical aspects of any industrial ecology study: the need to take a life-cycle approach, and the need to take a systems-based approach. However obvious and intuitive this may be in principle, in fact it is extremely difficult to implement for a number of reasons. Some are subtle, such as the lack of incentives for many economic actors to take a broader view of the relationship between their activities and natural systems. Others are more apparent, and quite familiar to practitioners: the lack of a prioritized and quantified system of values that can be translated into fully quantitative approaches; lack of data on emissions and their links to changes in environmental systems; lack of data and methods to determine the comprehensive and systemic effects of particular economic activities; lack of harmonized analytical methodologies and common assumptions underlying them; and lack of mechanisms for comprehensively evaluating risks, costs, and benefits of options.

It is all too easy to let these difficulties, and others, deter efforts to make immediate efforts to improve the environmental efficiency of existing industrial practices. This case study, which discusses a family of tools developed by the author and Tom Graedel while at AT&T, shows that this is unnecessary. In fact, workable methodologies for implementing the principles of industrial ecology as currently understood can be developed. In presenting this case study, the purpose is not to make the student of policy issues a technical expert in life-cycle assessment or Design for Environment methodologies. It is presented, rather, to provide an idea of what such approaches mean in practice using a system that has been at least partially tested in industrial systems.

The AT&T matrix system has two obvious attributes. First, it combines qualitative and quantitative factors, making the reliance on professional judgment explicit, yet providing a "figure of merit" that can be used, with appropriate caution, to compare alternatives. In this, it reflects the state of the art, which more resembles medicine than engineering at this point. A diagnosis that takes into account major effects over the life cycle is doable; a fully quantitative analysis based on mechanistic application of formula is not. A corollary of this is that the results of such assessments are useful for their educational and learning value, and to identify major areas where gains in environmental efficiency may be possible; use of such assessments for rigorous quantitative comparison between options is beyond the state of the art. Thus, for example, the life-cycle assessments that compare different kinds of diapers or plastic versus paper cups are interesting and informative exercises, but their results must be taken with a grain of salt. The common saying that you can predict the outcome of a life-cycle assessment by knowing who paid for it is overly cynical, but does indicate the sensitivity of such assessments to initial assumptions.

The second obvious characteristic of this matrix system is its aggregation of information at a high level, which makes it appear perhaps even oversimplistic. Undoubtedly much information is washed out in the data aggregation process, but the trade-offs are those of utility and also, to some extent, honesty: Neither the sufficiency of data nor available methodologies support a meaningful detailed approach yet. An experienced professional can use these matrices to obtain considerable insight into the major environmental impacts of a material, product and process, facility, or even an infrastructure component with a day or two of analysis. More complex methodologies, such as the life-cycle assessment system developed by SETAC (the Society of Environmental Toxicologists and Chemists), look at one material in one application, but may take six months or more, and cost over $100,000 U.S. to perform. Other systems, such as the EPS (Environmental Priority Strategies for Product Design) system developed by Volvo and the Swedish Government, are intermediate in complexity and offer significant promise for the future.

The point is not that either is right or wrong; rather, each has conditions under which its use is valuable, and others where it is inappropriate. Practitioners are just learning which is which, and are beginning to understand when to apply which tools and when to develop new ones. It is a time for experimentation, not doctrine. This has obvious policy implications: Experimentation by both private and public entities should be encouraged by appropriate policy tools (R&D credits, targeted public research efforts, public awards for significant achievements, and the like), but it is premature to embed any specific methodologies or results in regulations.

Evaluating the complex systems involved in industrial activities is a challenge in part because there are different levels of each system that must be connected by the analytical methodology, just as the operations are in practice. Accordingly, matrices have been developed for several levels of the systems hierarchy: each feeds into the next higher level of the system (refer to Figure 15.1). The base matrix, of course, evaluates materials over their life cycle. These results feed into product matrices (which include the processes by which the product is manufactured), which in turn support analyses at the facility level. Finally, depending on the purpose of the evaluation, one may go to the level of infrastructure assessment. The links among the matrices are not just functional, as illustrated in Figure 15.1, but also substantive. The facility matrix, for example, is based in part on the environmental preferability of the processes used to make the products inside it, which ties it directly to the results of the product/process matrices. If appropriate, therefore, this system can be used for a fairly comprehensive evaluation of firm operations across a number of scales. Many cases, however, will not

Matrix Evaluation System

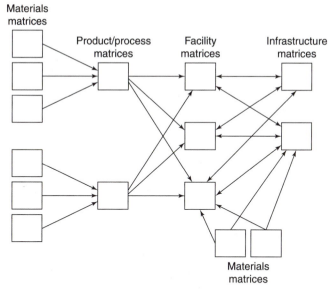

FIGURE 15.1   Overview of the AT&T Matrix System. Although the matrices can be used for individual material, process, product, facility or infrastructure assessments, they can also be combined to create a comprehensive assessment of an industrial system at many different levels. An additional level not illustrated in this diagram would be the evaluation of services for which the industrial system was the platform.

require application of all levels of the matrix system; where, for example, the only degrees of freedom that could be affected by the assessment involve product design, only process and product matrix assessments may be made. In fact, assessments within AT&T using this system have primarily focused on products.

A brief discussion of how a matrix evaluation is constructed will clarify the procedure. The materials matrix will be used as an example, as materials are obviously an important link between economic activity and impacts on underlying natural systems. A similar structure, including appropriate checklists, supports the other matrices in the system.

15.2   THE MATERIALS MATRIX SYSTEM

Just as industrial ecology urges a systems-based, life-cycle approach to product or facility design, it also requires that materials be subject to the same scope of evaluation. Indeed, a critical element in any evaluation of a process or product, or of built infrastructure for that matter, must be an evaluation of the materials with which it is initially constructed, how those materials are processed and used to create goods, and how the materials are reused, recycled, or disposed of. Thus, the material environmental evaluation matrix may be regarded as a foundation tool for the implementation of the principles of industrial ecology through the use of DFE methodologies. It attempts to capture a major class of social costs—specifically environmental and energy impacts

—implicit in material choices that may not be fully reflected in the economic data that forms the basis for most design decisions.

15.2.1 Characteristics of a Material Environmental Evaluation Matrix

In any effort of this sort, it is important to remember the nascent state of development of the field of industrial ecology and its implementation. Thus, for example, tools should not only be useful now, but also capable of being upgraded as knowledge and sophistication improve. What is presented here, therefore, represents an initial effort to define a DFE tool to evaluate the environmental impacts of materials: It will undoubtedly be considerably elaborated in the future. Nonetheless, it is equally important to begin now, with what is already known, as the potential for significant short-term improvements in environmental and economic efficiency are both real and achievable, so long as the theoretical best is not permitted to become the enemy of the actual good. Thus, any tool to support such an assessment should be robust: It should provide some basis to improve current practices even when data are sparse and uncertain, yet it should also easily reflect improved data and analysis as those become available. To the extent possible, it should include quantitative elements, without providing the false security or certainty that sometimes results from the inappropriate application of quantitative methods.

Another initial observation is that any methodology that is to be broadly applicable to materials must be generic and process oriented; it must ensure the right questions are asked and impacts considered, rather than being specific to the particular material or application. A process solvent or a material selected for the casing of a computer are vastly different in function and technology; yet with both it is appropriate and necessary to make the same basic evaluations. Accordingly, the matrix tool is designed to establish and support a generally applicable assessment process. In practice, of course, characteristics specific to each material, its initial and secondary processing, the intended application, and, if in a product, the use and fate of the product, and even associated co-evolved technologies with their suite of impacts, will come into play in performing the evaluation. Thus, for example, indium-based solders are in many cases incompatible with copper substrates, which are commonly used in the printed wiring boards used in electronics manufacture, and therefore a shift from lead to indium solders might require fundamental modification to the existing design of the boards. If so, the associated environmental impacts should be considered as part of the assessment.

Any material assessment tool should have several additional inherent characteristics:

1. It should support direct comparisons among material choices for various applications. Reflecting data limitations, these might initially be semi-qualitative, but should increasingly become quantitative (thus requiring a graceful ability to evolve toward quantification).

2. It should be usable and reasonably consistent across different assessment teams.

3. It should encompass all stages of the material's life cycle, and capture and identify the relevant (e.g., first order) environmental concerns.

4. It should be simple enough to permit relatively quick and inexpensive assessments to be made, yet robust enough to reflect continuing improvement in data and analytical techniques.

5. It should permit comparisons between materials (data sets) of widely differing quality, uncertainty, and completeness.
6. The results should be easy to comprehend and display. In the AT&T matrix system, two methods have been used: a target graph (shown below with data for AT&T products) and a line graph (shown below with results from made-up data).

15.2.2 Matrix Structure

The central feature of the AT&T matrix system is a 5 × 5 assessment matrix, one dimension of which is environmental concern and the other of which is life-cycle stage activities. The environmental concerns are generic across all matrices, corresponding to the five major classes of environmental concern: ecological and biological impacts, energy use, solid residues, liquid residues, and gaseous residues. While these categories are quite broad and others could no doubt be suggested, these are readily understood and reasonably comprehensive, in keeping with the practical intent of the system. Both local ecological impacts and (if applicable) loss of biodiversity could be comprehended in the first column, for example.

The life-cycle stages differ according to the focus of the matrix: materials (Figure 15.2), product or manufacturing process (Figure 15.3), facility (Figure 15.4), and infrastructure (Figure 15.5). Regarding materials, for example, the rows correspond to the generic life cycle of a particular material used in a particular application (modified slightly to fit a manufacturing example): the initial production and processing stage; the secondary production and processing stage; the application manufacturing stage; the application use stage; and the disposal or recycle stage.

In each case, an industrial ecology assessor studies the object of the assessment, identifies the relevant impacts, and assigns to each element of the matrix a rating from 0 (highest impact, a very negative evaluation) to 4 (lowest impact, an exemplary evaluation). The overall matrix rating ("MR") is then the sum of the matrix element values:

Material Environmental Evaluation Matrix

Lifecycle stage	Environmental concern				
	Ecological/ biological impacts	Energy use	Solid residues	Liquid residues	Gaseous residues
Initial production/ processing					
Secondary production/ processing					
Application: manufacturing stage					
Application: use stage					
Disposal, recycle					

FIGURE 15.2   Material Environmental Evaluation Matrix

Environmentally Responsible Product and Process Matrix					
	Environmental concern				
Lifecycle stage	Materials choice	Energy use	Solid residues	Liquid residues	Gaseous residues
Preproduction					
Product manufacture					
Product packaging and transport					
Product/process					
Recycling, disposal					

FIGURE 15.3   Environmentally Responsible Product and Process Matrix

Environmentally Responsible Facility Assessment Matrix					
	Environmental concern				
Lifecycle stage	Ecological/ biological impacts	Energy use	Solid residues	Liquid residues	Gaseous residues
Site selection, development, infrastructure					
Principal business activity-products					
Principal business activity-process					
Facility operations					
Facility refurbishment, closure, transfer					

FIGURE 15.4   Environmentally Responsible Facility Assessment Matrix

Environmentally Responsible Infrastructure System Matrix					
	Environmental concern				
Lifecycle stage	Ecological/ biological impacts	Energy use	Solid residues	Liquid residues	Gaseous residues
Site selection and preparation					
Infrastructure manufacture					
Infrastructure use					
Impact on ancillary systems					
Infrastructure end-of-life					

FIGURE 15.5   Environmentally Responsible Infrastructure System Matrix

$$MR = \frac{M_{i \cdot j}}{ij}$$

Because there are 25 matrix elements, a maximum rating for a specific assessment is 100.

In arriving at an individual matrix element assessment, or in offering advice to managers seeking to improve the rating of particular matrix element, the assessor uses detailed checklists and evaluation techniques. The use of such checklists is a common design approach and is familiar to engineers, which makes the matrix more easily assimilated than more exotic approaches. To provide an idea of what such checklists look like, an illustrative materials matrix checklist system is given at the end of this chapter. The breadth of issues that need to be considered and integrated even in something as apparently simple as choosing an environmentally efficient material for a specific application can be seen from this example.

Initially, use of such matrices will result in a semi-qualitative overall rating for each cell. With experience and more adequate data, however, quantitative techniques can begin to be implemented for individual cells as appropriate; the checklists can be made more rigorous. The procedure accommodates, however, the interim situation—which could last for a very long time—in which certain matrix assessments are more easily quantified than others, and in which more data are available for certain aspects of materials, products, facilities, and infrastructure than others. Under such circumstances, ranking by order of merit is still possible using this methodology, but must be done with due appreciation for the judgmental factors involved. Thus, a difference of a few points in the figures of merit of different options probably is immaterial, while a difference of 15 or 20 may well indicate a meaningful difference in environmental efficiency.

The assignment of a discrete value from zero to four for each matrix element implicitly assumes that the DFE implications of each element are equally important. An option for slightly increasing the complexity of the assessment (but perhaps increasing its utility as well) is to reflect detailed environmental impact information by applying weighting factors to the matrix elements. For example, if energy used in initial production of the material (for example, mining and initial processing of metal ore) were judged because of global climate change concerns to outweigh the localized impacts of liquid residues, the "energy use" column weighting could be enhanced, and that of the "liquid residue" column correspondingly decreased. Any such weighting should be reasonably robust across different materials, however, because comparisons of different material/application options would require the use of identical weighting factors. To do otherwise would introduce different underlying value decisions into an otherwise relatively transparent procedure.

This system is deliberately semi-quantitative to respond to the conundrum that has often bedeviled attempts to develop workable DFE and life-cycle assessment (LCA) tools. On the one hand, it is extremely difficult—many professionals would say impossible—to quantify the impacts of even those environmental releases and effects that can be inventoried. How should one quantitatively evaluate, for example, a trade-off between using a substance with a highly uncertain potential for human carcinogenicity, and possible loss of biodiversity? On the other hand, quantitative systems are a prerequisite for diffusion of DFE methodologies and concepts throughout industry, especially if modifications to business planning and highly computerized design processes are desired. The matrix system thus explicitly relies on the professional judgment of industrial ecologists, while allowing for standardization of dimensions through common checklists as the state-of-the-art advances and experience is gained; yet it produces a quantitative evaluation as well. It therefore provides an easily used management and operational tool.

Once an assessment has been done, the results can be used in a number of ways. In the case of materials, for example, the results can help rank different materials in various applications, thereby supporting the selection of the most environmentally efficient material for the application, within other technological or economic constraints. The latter qualification is worth noting: These tools increase environmental efficiency for private firms, but cannot compensate for externalities and other inefficiencies within the broader economy. If energy, agricultural products, or environmental control technology is systematically underpriced or subsidized, for example, it will generate economic and environmental inefficiencies that a private firm, if it is to remain competitive, can only reduce marginally if at all. It must be remembered that environmental data are only one input into the design process, and that the usual design constraints, such as cost, performance, time-to-market, and material availability and physical properties, continue to hold as well.

Alternatively, the assessment can be used for an improvement analysis of a specific material in a particular application. This function can be facilitated by displaying data in a line graph, which visually identifies issues that should be singled out for improvement. Such a graph is constructed by plotting each element of the matrix as shown in Figure 15.6, constructed using arbitrary data. Environmentally preferable characteristics are clustered toward the bottom; less desirable aspects are higher. Thus, given the data in Figure 15.6, one would initially focus on improving the production of gaseous residues during the manufacturing stage (element 3,3), and the ecological and biological impacts associated with disposal and recycle of the material (element 5,1). Moreover, sets of line graphs—illustrating, for example, different materials in the same application—can be directly contrasted, providing an easily understood basis for decision.

One can also construct a *target plot*. Figure 15.7, for example, is a target plot created by Graedel et al. based on actual data generated by evaluating over 15 AT&T electronics products using the products matrix illustrated in Figure 14.3. Points clustered near the center indicate environmentally preferable results; points nearer to the outside indicate opportunities for improvement. Thus, the result for matrix cell (3,3) reflects the fact that the recyclability of packaging needs to be improved, and cell (5,3) indicates that the existing designs offer little opportunity to reuse materials.

15.2.3   The Materials Matrix Checklist

It is useful for any individual interested in policy in this area to appreciate the scope, and numerous difficulties, involved in assessment of environmental impacts of various activities of the firm. Accordingly, this section presents a sample of possible items to be included in the checklist for each matrix cell, using the materials matrix as an example.

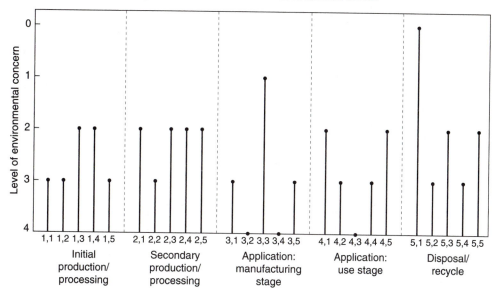

FIGURE 15.6   Line graph of material environmental evaluation matrix analysis results, using arbitrary data. Such a display makes it easy to identify opportunities for improvements in environmental efficiency, in this case, element 3,3 (gaseous residues in manufacturing lifecycles stage), and element 5,1 (biological impacts at disposal /recyclying stage).
    Source: Based on Allenby, B.R., A design for environment methodology for evaluating materials, Total Quality Environmental Management, Summer 1996, pp. 69-84.

Different materials and applications may require different checklists because they will involve different impacts and technologies. This checklist, therefore, should be regarded as illustrative rather than definitive.

Material Environmental Evaluation Matrix Element 1,1

Life-cycle Stage: Initial Production/Processing
Environmental Concern: Ecological/Biological Impacts

- Is the location where the initial production or processing of the material occurs ecologically sensitive?
- Do the techniques and technologies used in initial production or processing meet world standards?
- If new infrastructure must be created, are plans in place to minimize any resultant impact on biota?
- Does the initial production or processing of this material create any other material streams (e.g., co-products or by-products) that have problematic impacts on biota at any point in their life cycles?
- Does the production of this material have any potential impacts on biodiversity?

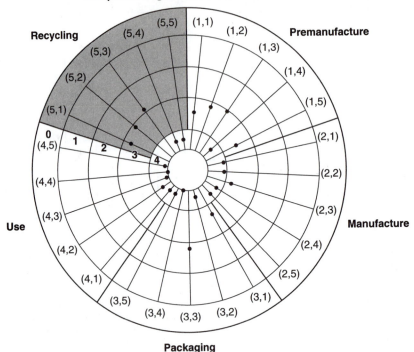

FIGURE 15.7   Target plot of material environmental evaluation matrix analysis results, based on aggregated AT&T data. Recyclability of packaging (element 3,3) and material reuse (element 5.3) are identified as areas for possible improvement.
   Source: Graedel, T.E., P.R Comrie and J.C. Schutowski, Green Product Design, AT&T Technical Journal 74(6): 17-25.

Material Environmental Evaluation Matrix Element 1,2

Life-cycle Stage: Initial Production/Processing
Environmental Concern: Energy Use

- Does initial production or processing of this material involve significant amounts of energy?
- Does initial production or processing of this material involve significant use of transportation resources?
- Does the preparation of initial production or processing facilities or infrastructure involve significant amounts of energy?
- To the extent possible, is required energy generated using environmentally appropriate technologies?
- To the extent possible, is all production and processing equipment energy efficient (e.g., uses variable speed motors)?

Material Environmental Evaluation Matrix Element 1,3

Life-cycle Stage: Initial Production/Processing
Environmental Concern: Gaseous Residues

- Are initial production and processing activities conducted such that only minimal streams of gaseous residues are produced?
- Have plans been made to minimize the impacts of any gaseous residues that are produced?
- Have the initial production and processing activities been designed to reduce the toxicity of any gaseous residues to the extent possible?
- Have any gaseous residue streams produced been designed to optimize any economic value they might have?

Material Environmental Evaluation Matrix Element 1,4

Life-Cycle Stage: Initial Production/Processing
Environmental Concern: Aqueous Residues

- Are initial production and processing activities conducted such that only minimal streams of aqueous residues are produced?
- Have plans been made to minimize the impacts of any aqueous residues that are produced?
- Have the initial production and processing activities been designed to reduce the toxicity of any aqueous residues to the extent possible?
- Have any aqueous residue streams produced been designed to optimize any economic value they might have?

Material Environmental Evaluation Matrix Element 1,5

Life-Cycle Stage: Initial Production/Processing
Environmental Concern: Solid Residues

- Are initial production and processing activities conducted such that only minimal streams of solid residues are produced?
- Have plans been made to minimize the impacts of any solid residues that are produced?
- Have the initial production and processing activities been designed to reduce the toxicity of any solid residues to the extent possible?
- Have any solid residue streams produced been designed to optimize any economic value they might have?

Material Environmental Evaluation Matrix Element 2,1

Life-Cycle Stage: Secondary Production/Processing
Environmental Concern: Ecological/Biological Impacts

- Is the location where the secondary production or processing of the material occurs ecologically sensitive?

- Do the techniques and technologies used in secondary production or processing meet world standards?
- If new infrastructure must be created, are plans in place to minimize any resultant impact on biota?
- Does the secondary production or processing of this material create any other material streams (e.g., co-products or by-products) that have problematic impacts on biota at any point in their life cycles?
- Does the secondary production or processing of this material have any potential impacts on biodiversity?

Material Environmental Evaluation Matrix Element 2,2

Life-Cycle Stage: Secondary Production/Processing
Environmental Concern: Energy Use

- Does secondary production or processing of this material involve significant amounts of energy?
- Does secondary production or processing of this material involve significant use of transportation resources?
- Does the preparation of secondary production or processing facilities or infrastructure involve significant amounts of energy?
- To the extent possible, is required energy generated using environmentally appropriate technologies?
- To the extent possible, is all production and processing equipment energy efficient (e.g., uses variable speed motors)?

Material Environmental Evaluation Matrix Element 2,3

Life-Cycle Stage: Secondary Production/Processing
Environmental Concern: Gaseous Residues

- Are secondary production or processing activities conducted such that only minimal streams of gaseous residues are produced?
- Have plans been made to minimize the impacts of any gaseous residues that are produced?
- Have the secondary production or processing activities been designed to reduce the toxicity of any gaseous residues to the extent possible?
- Have any gaseous residue streams produced been designed to optimize any economic value they might have?

Material Environmental Evaluation Matrix Element 2,4

Life-Cycle Stage: Secondary Production/Processing
Environmental Concern: Aqueous Residues

- Are secondary production or processing activities conducted such that only minimal streams of aqueous residues are produced?

- Have plans been made to minimize the impacts of any aqueous residues that are produced?
- Have the secondary production or processing activities been designed to reduce the toxicity of any aqueous residues to the extent possible?
- Have any aqueous residues produced been designed to optimize any economic value they might have?

Material Environmental Evaluation Matrix Element 2,5

Life-Cycle Stage: Secondary Production/Processing
Environmental Concern: Solid Residues

- Are secondary production or processing activities conducted such that only minimal streams of solid residues are produced?
- Have plans been made to minimize the impacts of any solid residues that are produced?
- Have the secondary production or processing activities been designed to reduce the toxicity of any solid residues to the extent possible?
- Have any solid residues produced been designed to optimize any economic value they might have?

Material Environmental Evaluation Matrix Element 3,1

Life-Cycle Stage: Applications: Manufacturing Stage
Environmental Concern: Ecological/Biological Impacts

- Do all outputs from the manufacturing stage, including residue streams, have high ratings as environmentally responsible products or materials?
- Does the manufacturing activity, especially that associated with this material, use recycled or refurbished inputs?
- Does the manufacturing process meet world standards for worker safety; exposures of humans, other organisms, or biological or human communities; and safe material-handling procedures?
- Are transportation and packaging impacts ancillary to the manufacturing activity minimized?

Material Environmental Evaluation Matrix Element 3,2

Life-Cycle Stage: Applications: Manufacturing Stage
Environmental Concern: Energy Use

- Does the manufacturing process associated with this material minimize the use of energy-intensive process steps such as high heating differentials, heavy motors, extensive cooling, and so on?
- Does the manufacturing process associated with this material minimize the use of energy-intensive evaluation steps such as testing in a heated chamber?
- Do the relevant manufacturing activities use co-generation, heat exchange, and other techniques for utilizing otherwise wasted energy?

Material Environmental Evaluation Matrix Element 3,3

Life-Cycle Stage: Applications: Manufacturing Stage
Environmental Concern: Gaseous Residues

- Are CFCs, HCFCs, or chlorinated solvents used in the applicable manufacturing activities, and, if so, have alternatives been thoroughly investigated?
- Are any greenhouse gases used or generated, and, if so, have alternatives been thoroughly investigated?
- Are gaseous residues designed for minimum toxicity, minimum volume, and optimum reuse?
- Are the applicable manufacturing activities designed to utilize the maximum amount of recycled gaseous materials?

Material Environmental Evaluation Matrix Element 3,4

Life-Cycle Stage: Applications: Manufacturing Stage
Environmental Concern: Aqueous Residuals

- Are aqueous residues and process inputs designed for minimum toxicity, minimum volume, and optimum reuse?
- If solvents or oils are used in any appropriate manufacturing process, is their use minimized and have substitutes been investigated?
- Are all water treatment and management systems associated with liquid residual flows up to world standard?
- Is any transportation of liquid residuals minimized?

Material Environmental Evaluation Matrix Element 3,5

Life-Cycle Stage: Applications: Manufacturing Stage
Environmental Concern: Solid Residuals

- Are all packaging and transportation imports associated with solid residuals minimized?
- Are all solid residues and inputs designed for minimum toxicity, minimum volume, and optimum reuse?
- Are all treatment, handling and management systems associated with solid residuals world class?

Material Environmental Evaluation Matrix Element 4,1

Life-Cycle Stage: Applications: Use Stage
Environmental Concern: Ecological/Biological Impact

- Does this material cause any biological or ecological impact when the product, process, or service within which it is embedded is used?
- Is the material intentionally or incidentally dispersed during application use in such a way as to potentially impact biological or ecological systems?

- Does the material cause any worker, consumer, or other human health impact during application use?

Material Environmental Evaluation Matrix Element 4,2

Life-Cycle Stage: Applications: Use Stage
Environmental Concern: Energy Use

- Does the material contribute to energy minimization during application use?
- Does the material in application use contribute to energy efficiency in any associated products or technologies?
- Does the weight of the material contribute to energy (in)efficiency during application use?

Material Environmental Evaluation Matrix Element 4,3

Life-Cycle Stage: Applications: Use Stage
Environmental Concern: Gaseous Residues

- Does the material contribute to any gaseous residues during application use?
- If so, has the toxicity and the volume of residual been minimized?
- If intentional dissipative gaseous emissions occur, have non-dissipative alternatives been considered?

Material Environmental Evaluation Matrix Element 4,4

Life-Cycle Stage: Applications: Use Stage
Environmental Concern: Aqueous Residues

- Does the material contribute to any aqueous residues during application use?
- If so, has the toxicity and the volume of residual been minimized?
- If intentional dissipative aqueous emissions occur, have non-dissipative alternatives been considered?

Material Environmental Evaluation Matrix Element 4,5

Life-Cycle Stage: Applications: Use Stage
Environmental Concern: Solid Residues

- Does the material contribute to any solid residues during application use?
- If so, has the toxicity and the volume of residual been minimized?
- If intentional dissipative solid emissions occur, have non-dissipative alternatives been considered?
- Has packaging associated with this material in its application use been minimized and designed for recycling?

Material Environmental Evaluation Matrix Element 5,1

Life-Cycle Stage: Disposal/Recycle
Environmental Concern: Ecological/Biological Impacts

- Are the facilities associated with disposal or recycling of the material designed for minimal ecological, biological, and worker and other human health impacts?
- Does the disposal or recycling of this material create any other problematic material streams?
- Is the material in this application inherently recyclable, and is it economically and environmentally efficient to do so?
- If so, is it actually recycled?
- Are the transportation and other infrastructure systems necessary for disposal or recycling designed to produce minimal ecological, biological, and worker and other human health impacts?

Material Environmental Evaluation Matrix Element 5,2

Life-Cycle Stage: Disposal/Recycle
Environmental Concern: Energy Use

- Are the facilities associated with disposal or recycling of the material designed for, and maintained to achieve, optimum energy efficiency?
- Are the transportation and other infrastructure systems necessary for disposal or recycling designed and maintained to obtain optimum energy efficiency?
- Is it environmentally preferable to recycle the material for its energy content (e.g., burn for co-generation)?

Material Environmental Evaluation Matrix Element 5,3

Life-Cycle Stage: Disposal/Recycle
Environmental Concern: Gaseous Residues

- Have the volume and toxicity of any gaseous residues produced by disposing of or recycling the material been minimized?
- Is the technology used to dispose of or recycle the material world class in terms of environmental and economic efficiency?
- Are all gaseous emissions during disposal or recycling activities, if any, handled and managed appropriately?
- Are all infrastructure systems required for disposal or recycling, including transportation and packaging, designed to minimize the volume and toxicity of gaseous residues?

Material Environmental Evaluation Matrix Element 5,4

Life-Cycle Stage: Disposal/Recycle
Environmental Concern: Aqueous Residues

- Have the volume and toxicity of any aqueous residues produced by disposing of, or recycling, the material been minimized?
- Is the technology used to dispose of or recycle the material world class in terms of environmental and economic efficiency?
- Are all aqueous emissions during disposal or recycling activities, if any, handled and managed appropriately?
- Are all infrastructure systems required for disposal or recycling, including transportation and packaging, designed to minimize the volume and toxicity of aqueous residues?

Material Environmental Evaluation Matrix Element 5,5

Life-Cycle Stage: Disposal/Recycle
Environmental Concern: Solid Residues

- Have the volume and toxicity of any solid residues produced by disposing of or recycling the material been minimized?
- Is the technology used to dispose of or recycle the material world class in terms of environmental and economic efficiency?
- Are all solid emissions during disposal or recycling activities, if any, handled and managed appropriately?
- Are all infrastructure systems required for disposal or recycling, including transportation and packaging, designed to minimize the volume and toxicity of solid residues?

REFERENCES

AEA (American Electronics Association). *The Hows and Whys of Design for the Environment.* Washington, D.C.: 1993.

Allenby, B.R. "Design for Environment: A tool whose time has come." *SSA Journal* September, 5–9: 1991.

Allenby, B.R. "A Design for Environment methodology for evaluating materials." *Total Quality Environmental Management* 5(4):69–84. Summer 1996.

Allenby, B.R. and T.E. Graedel. "Defining the environmentally responsible facility," Third Annual National Academy of Engineering Industrial Ecology Workshop, June, Woods Hole, MA: 1994 (to be published in a forthcoming volume by National Academy Press, Washington, D.C.)

Allenby, B.R. and D.J. Richards, eds. *The Greening of Industrial Ecosystems.* Washington, DC, National Academy Press: 1994.

Ayres, R.U. and V.E. Simonis, eds. *Industrial Metabolism.* Tokyo, United Nations University Press: 1994

Graedel, T.E. and B.R. Allenby. *Industrial Ecology.* Upper Saddle River, NJ, Prentice-Hall: 1995.

Graedel, T.E., P.R. Comrie, and J. C. Sekutowski. "Green product design." *AT&T Technical Journal* 74(6):17–25 November/December 1995.

Socolow, R., C. Andrews, F. Berkhout, and V. Thomas, eds. *Industrial Ecology and Global Change.* Cambridge, Cambridge University Press: 1994.

Turner II, B.L., W.C. Clark, R.W. Kates, J.F. Richards, J.T. Mathews, and W.B. Meyer, eds. *The Earth as Transformed by Human Action.* Cambridge, Cambridge University Press: 1990.

EXERCISES

1. Select three items from your kitchen: a simple product (one whose function depends primarily on its material content, such as a soap or detergent), a food item, and a complex product (one whose function depends primarily on design, such as a coffee maker or a refrigerator). Using the product matrix (Figure 14.3), perform an analysis of each product.

   a.   How do the Figures of Merit (FOM) for each product compare?

   b.   What are the most critical assumptions that you had to make in deriving the FOM, and how sensitive is the FOM to that assumption (that is, if you change your assumption, does it change the resulting FOM by a significant amount)?

   c.   Is the same life-cycle stage the most problematic for each product? If not, what does the difference tell you?

   d.   Using your results, suggest a reasonably practical policy that in each case would help improve the environmental efficiency of each product. Are there any commonalities to the policies you would recommend?

2. You are an automotive designer. You have a choice of two materials for the bonnet (front hood) of a car: steel or a carbon/plastic composite. The steel is heavier than the composite, takes more energy to make, but, unlike the composite, can be (and is) recycled at the end of the automobile's useful life.

   a.   Using the material matrix and the checklist provided above, compare the two options.

   b.   What are the major impacts of each choice? What assumptions did you have to make when comparing the two different materials in this application?

   c.   What policy implications does this simple exercise suggest to you?

   d.   Based on your analysis, what are the strengths and weaknesses of the matrix methodology?

CHAPTER 16

Is the Private Firm Compatible with a Sustainable World?

This text speaks extensively of the need for continuing evolution of new, environmentally preferable technology and technological systems. Technology development and deployment is, by-and-large, the province of private firms in a market economy. Accordingly, it is necessary to consider the institution of the private firm, and its effectiveness and appropriateness as a vehicle for supporting the achievement of sustainability. Doing so in an unusual context—that of the private firm as an agent in a complex system—provides interesting insights into a fundamental question of economic structure in a sustainable world: Can the private firm as currently constituted continue in an environmentally constrained world? And, if so, what additional social or cultural structures, if any, are required?

Unlike the three other case studies in this Part, this one is fairly conceptual, and is designed to illustrate one of a class of fundamental questions about human institutions, technology, and modern trends that an industrial ecology-based approach to policy may raise. It provides no ready answers, which, although perhaps frustrating, accords with the difficulty of the issues involved and the reality of a rapidly changing world.

16.1 INTRODUCTION TO THE CASE

A number of trends and proposals reflecting rising global environmental concern implicitly suggest an evolution in private, for-profit firms away from profit-seeking behavior toward more socially comprehensive goals. Indeed, some firms are even calling themselves "sustainable firms," although the substantive content of this assertion is not clear. It is not apparent, however, that all of the implications of such an evolution have been considered. In considering this issue, two polar positions can be identified: (1) that firms should be modified to become the vehicles of sustainability, internalizing values, norms, and behavior to become stewards of the environment and associated social values; and (2) that sustainability can best be achieved by modifying the boundary conditions within which private corporations operate, rather than by trying to change them into organizations that reflect social goals beyond profit-seeking. If the first position is taken, substantial evolution in culture and law is implied. If the second is taken, it would suggest that the social and legal responsibilities of corporations not be inadvertently and unintentionally expanded beyond that currently established. Rather, wise policy should create incentives for corporations to behave in ways supporting the achievement of sustainable societies. Note that this requires that authority to create and enforce such policies rest with some institution other than the firm itself.

This case study should be considered in the context of broader trends that suggest accelerating shifts of power and responsibilities among political and economic entities at all levels. The post Cold War political environment has seen a number of instances of devolution of power from existing national states to internal regions or political groups, either violent, as in the case of the breakup of Yugoslovia, or still peaceful, as in the case of Quebec in Canada or the breakup of Czechoslovakia.

Even where national integrity is assured, the last decade has seen shifts of power from states to lesser internal jurisdictions, as in the United States. Taxation authority, for example, is increasingly being shifted to sub-state jurisdictions, and legal responsibilities in areas such as environmental regulation are increasingly centered at the state and local, not the national, level. Conversely, the European Union, attempting to fashion a super-national organization from member European states, is an interesting, if somewhat contentious, example of power perhaps devolving toward political entities larger than the state. In the Americas, the North American Free Trade Agreement (NAFTA) is a similar case. While it purports to be simply a free-trade agreement, the blurring of NAFTA border areas in terms of culture, population migration, metropolitan areas, and industrial activity suggests that a more fundamental pattern of super-national regionalization is actually occurring.

Political devolution of power away from the national state is mirrored in the economic sphere. There is thus a relatively large and growing literature on how the degrees of freedom of the state in many critical areas—for example, control of national financial activities, control of information flows, and control of technology—are increasingly limited by a global economic system structurally dominated by large transnational corporations. The private role in international development is expanding significantly compared to the traditional, multinational donors. Increasingly, for example, support for Asian and Latin American growth comes not from traditional government and international aid programs, but from private foreign direct investment (FDI). Moreover, internally, privatization of nationalized firms in "sensitive" sectors such as energy and telecommunications is accelerating in developing and developed countries alike: Chile has even privatized their future social security/pension funds. A parallel development of the power and roles of nonprofit, non-governmental organizations (NGOs), especially in the environmental area, balances the evolution of the for-profit corporate sector.

In one sense, the issues raised by these activities are not unique to the modern age. The "company towns" of the late 19th century in the United States raised similar issues of paternalism, the degree to which private firms should be "socially responsible," the appropriate division between private and public interests, and potential exploitation of workers and their families by private firms, albeit at a much reduced scale. It is apparent, however, that yet another fundamental realignment of functions between the national state and the private sector is underway. This is an important issue for this case study to the extent that it suggests that national states may have less capability in the future to bound the behavior of private firms. If states, or some other as yet unrecognized social structure (some have suggested international NGOs), cannot perform this boundary setting function, the possibility arises that less regulated for-profit firms may well be incompatible with a sustainable economy.

This case views private firms in two, not incompatible ways. One is as the familiar standard economic unit of activity. The second is as an agent in a complex system, as those terms are defined in Chapter 8. In this regard, the term "boundary conditions" as used in this case study can be thought of as the set of factors that change the external environment within which the firm operates, but do not explicitly change the firm itself. In this sense, a prescriptive regulation regarding a specific activity—mandating use of a given scrubber technology, for example—would not be considered a boundary condition; as a mandatory legal requirement, it is a direct prescription of an aspect of the firm's internal state, not an incentive for a behavior change. A requirement that a firm release information about its emissions, however, is a regulation affecting boundary conditions. Its impact arises not from the minimal information requirement, but from the incentives that such information, when released to the public, creates for the firm (and to which the firm can respond in many ways). As this example suggests, most policies will be a combination of changes in boundary conditions and targeted intervention; in most cases, however, one or the other aspect will clearly dominate. Generally, a boundary condition leaves the firm, as an active agent, with a number of degrees of freedom, which encourages competition and evolution within the system as a whole. A mandate, on the other hand, is incompatible with evolution and tends to be economically—and in many cases, environmentally—inefficient.

With this as background, the central question raised by this case study can then be stated: What forms of corporate structure are current trends tending to evolve, and is this the form of corporate structure most likely to contribute to the evolution of economic structure and behavior that is dynamically stable over the long term in an environmentally constrained world? The purpose of this case study is not to answer that question definitively—it is doubtful that enough is known yet in many ways to do so—but to identify the trends that might drive redefinition of the corporation, to lay the groundwork for a rational discussion of the desirability of any such redefinition, and to illustrate the policy issues and alternatives such an evolution might imply.

16.2 THE PRIVATE FIRM

All economic systems are characterized by processes of production, but the existence of private firms tends to characterize only those systems commonly referred to as *market economies*, in which the means of production are by-and-large privately owned, and the profit motive dominates. Moreover, any discussion of the behavior of private firms must differentiate between the private enterprise system taken as a whole, the firms within that system, and the individuals constituting the firm. Policy options are very different, for example, if the private enterprise system as a whole is generating too much waste, as opposed to an individual firm. In one case, comprehensive legislation may be required, while in the other targeted legal action may be appropriate. An individual who chooses to contribute a substantial part of her salary to charity functions is acting in a quite different capacity, and will be treated quite differently by the legal system, than if she, as a manager of a firm, decides to do the same thing with the firm's profits.

Private firms are pivotal economic agents in any modern economy: They reflect and create the cultures and economies within which they function. They also are creatures of law: created and defined by law, they may be modified by law. Firms are not static entities, however, but are remarkably protean, especially in times of rapid economic change. Recently, for example, a good deal of ink has dealt with "virtual firms," contractual entities, usually with a limited, project-specific focus, that rapidly form and dissolve. On a higher level of the hierarchy of economic systems, some interesting work is being done on self-organizing and loosely linked groups of firms that form frequently successful and innovative industrial districts in certain regions, such as Silicon Valley in the United States, or the collection of textile firms near Florence, Italy, known as "the Third Italy." Such combinations and collaborations can be quite successful economically, and, along with other "organizational experiments," may be expected to proliferate in a period of rapidly changing economic conditions. In her study of Silicon Valley and Route 128, for example, Annalee Saxenian notes that:

> Silicon Valley has a regional network-based industrial system that promotes collective learning and flexible adjustment among specialist producers of a complex of related technologies. The region's dense social networks and open labor markets encourage experimentation and entrepreneurship. Companies compete intensely while at the same time learning from one another about changing markets and technologies through informal communication and collaborative practices; and loosely linked team structures encourage horizontal communication among firm divisions and with outside suppliers and customers. The functional boundaries within firms are porous in a network system, as are the boundaries between firms themselves and between firms and local institutions such as trade associations and universities.

A less dramatic development that nevertheless shifts the boundaries of firms is the rapid increase in collaborative arrangements and institutions, where competitors formally join together to address issues of common interest. While trade associations and industry lobbying groups have been around for many decades, these new institutions are characterized by a substantive focus on critical business issues, and they usually involve targeted research and development, frequently bringing in additional parties that might range from academic researchers to national laboratories. Examples from the United States include the two electronics research and development consortia, the Microelectronics and Computer Technology Corporation, and Sematech; the Industry Cooperative for Ozone Layer Protection (ICOLP), the electronics industry group that led the sector away from CFC use; and such automobile industry collaborations as USCAR and the Partnership for a New Generation of Vehicles (PNGV). Figure 16.1 shows the dramatic growth in such collaborative institutions in the highly competitive automotive sector in the United States.

As these developments illustrate, an argument can be made that the private firm as the primary economic entity is undergoing as fundamental a redefinition as the national state, the primary political entity, and with just as much uncertainty about the final outcome of the process. Although this makes establishing the hierarchical level of the prototypic "firm" in the economic system a little less straightforward, the legal and social forces at issue—the profit motive, social concern about environmental impacts of economic activity—are generally applicable to many of these more exotic corporate forms.

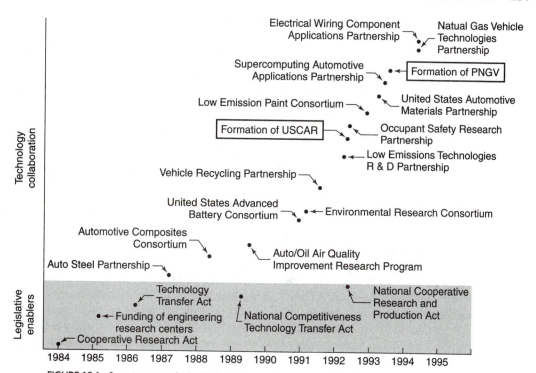

FIGURE 16.1 In many sectors in the United States, such as the automobile industry, consortia and collaborative initiatives have multiplied over the past few years. Evolution of these institutions has required legislation in many cases.

The fundamental purpose of a for-profit corporation in a free market economy is to make money for its owners, that is, the shareholders, a point made vigorously by Milton Friedman: "Few trends could so thoroughly undermine the very foundations of our free society as the acceptance by corporate officials of a social responsibility other than to make as much money for their stockholders as possible. This is a fundamentally subversive doctrine."

It is, of course, now generally recognized that firms to a greater or lesser extent reflect the interests of many stakeholders, including their employees, their managers, their customers and, increasingly, different governments in a global economy. Corporate charitable contributions, for example, are generally legal if they are reasonable, and the firm's directors have broad discretion to take such actions if they arguably help protect the corporation's interests. Additionally, firms from cultures less individualistic and adversarial than that of the U.S., such as Europe or, especially, Japan, tend to display a somewhat greater propensity to internalize the public interest, at least selectively. Inability to optimize one goal (or wisdom in a complex and rapidly changing environment) may lead firms to try to "satisfice"—meet minimum levels of performance—for several goals, rather than focus on one. Additionally, firms, like any agent in a complex system, must evolve in structure to reflect changing boundary conditions, such as, for example, shifting from an authoritarian management model and hierarchical structure appropriate to mass-production manufacturing to a flatter, specialist-based, information rich organization to reflect changes in technology and global market conditions.

Adaptation is not altruism, however, at least not at this point. Rothenberg has noted some of the difficulties inherent in relying on altruistic behavior on the part of private firms to resolve pollution issues: "The prospects rest on the willingness of equity holders to sacrifice income and other goals; on the ability of managers to formulate consistent rules of behavior; on the amassing of very sizable amounts of difficult and intrinsically ambiguous information beyond what is required for each firm's own operations; and on the ability and desire of numbers of altruistic firms to coordinate their efforts." To that still valid list of concerns can be added the obvious problem of achieving the level of public and political trust in private firms that reliance on altruism would require. It thus remains true that, broadly speaking, seeking profits remains the primary goal of the private corporation, and that the "public interest" theory of the firm, which was especially popular in the late 1960s, has not been accepted either generally or in the specific case of the environment.

The private corporate enterprise is such an intrinsic part of the modern capitalist economy that few realize its relative youth. It is possible to trace the antecedents of the corporation back to the medieval merchant guild systems, or, more recently, to trading companies enjoying monopolies granted under royal charter, such as the British East India Company. However, the advent of the truly modern firm awaited the development of *general incorporation laws*, under which any entity meeting statutorily defined criteria was able to incorporate, rather than the special grants of privilege that had hitherto prevailed. In the United States, the first of these laws was not passed until 1811, by New York State. As Rosenberg and Birdzell point out, the pattern subsequently established in Western economies—a complex network of independent firms, frequently competing on the basis of technological and scientific creativity, with successful innovation rewarded in the marketplace—became the basis for modern, materially successful economies. Thus, the modern corporation appeared at a certain stage in the development of the Industrial Revolution in large part, arguably, because such a construct was necessary for the continued evolution, with concomitant increasing complexity, of the industrial economies characterizing the modern national state. Indeed, it can also be argued that such entities as "virtual firms" and the Silicon Valley "district firms" represent a continuing evolution of the firm into the more complex and flexible entities necessary for a still more complex, post-industrial, service-oriented information age global economy.

16.3   THE FIRM AS AGENT IN A COMPLEX SYSTEM

As with most human inventions, the institution of the firm can be seen in many ways. The more traditional one is sketched above, but there is another that may be useful in increasing the understanding of the role of private firms in supporting the achievement of sustainability. In this view, the firm is viewed as an *agent*, an actor capable of adaptation at a given level or hierarchy within a complex system, which in this case is simply the global economy. The agent (firm) is itself a system at its level of the hierarchy, and is capable of limited foresight.

Let us then reconsider the question initially posed in this case study. It is at least a defensible hypothesis that sustainability is indeed an emergent characteristic of a properly organized and bounded complex system (the economy), and cannot therefore be defined *a priori* except in general terms. Moreover, it is also apparent that firms, as agents with limited foresight in this economy, function at a critical level in the system's hierarchy, that is, their behavior may be anticipated to have some impact on whether the system as a whole exhibits the emergent characteristic of sustainability. The policy debate can then be framed with two questions:

1. What form of agent at the hierarchical level of the firm is most likely to lead to the evolution of the economy, viewed as a complex system, in such a way that it exhibits the property of being sustainable over time?
2. What boundary conditions on the behavior of such agents—in other words, what policy structures—will be most effective in evolving the global economic system so that it will display the emergent characteristic of sustainability, given that such a state will be difficult to define *a priori*?

One interesting conceptual point should be noted in passing. This formulation assumes that, through policies or individual initiatives, society is capable of structuring firms within the context of a complex economy in such a way as to migrate toward a long-term, stable carrying capacity, or sustainability. It is not, however, intuitively apparent that individuals in a complex system, alone or in concert, are capable of understanding that system (or acting on that understanding) so as to deliberately structure it to lead to the emergence of the desirable self-organization. Nonetheless, achieving sustainability while avoiding serious cultural, social, economic, and population perturbations may depend on society's ability not just to understand complex systems (local, regional, and the global economies in both their, physical and financial dimensions, as well as underlying natural systems), but to understand how and where it is possible, in the real world of politics and interest groups, to interfere with and change them constructively. This may require a degree of understanding of ourselves and the systems within which we are embedded, and a political and cultural maturity necessary to act on that understanding, which is unprecedented and perhaps unlikely.

16.4 RECENT TRENDS

As yet, there has been little explicit recognition that inherent in the integration of technology and environment are a number of implicit drivers for the redefinition of the corporation as it is now known. Major ones include the following.

16.4.1 The Fundamental Conflict Between Uncontrolled Growth and an Environmentally Constrained World

If this conflict, emphasized by ecological economists among others, is valid, it implies a further fundamental conflict between the traditional goals of the firm—material wealth creation, maximization of value, and growth of value over time for shareholders —and achievement of a sustainable state. There are, however, conditions under which individual firms might grow even if unlimited economic growth, as opposed to economic development, is incompatible with sustainability. Some firms, for example, may continue to grow within a sector at the expense of less efficient firms, even if the sector itself remains the same size or even shrinks. Moreover, as sustainability is approached, some economic sectors will undoubtedly grow at the expense of others as the economy restructures itself. Thus, for example, it is at least plausible that sustainability may require the substitution of information and intellectual resources for material and energy inputs during manufacture or use, leading to an expansion of the electronics and telecommunications sector in the economy as a whole. Some forms of service, such as creating appropriate software applications, may contribute to dematerialization of the economy; firms providing such products could, and should, increase their value as sustainability is approached.

It is, in fact, only when the growth orientation of corporations results in the inappropriate growth of the economy taken as a whole—in, for example, capital stock, energy consumption, flows of materials, or waste generation—that sustainability may be challenged. It is thus possible that corporations could retain a strict growth incentive while society creates boundary conditions that preclude the unsustainable growth of the economy. For example, the availability of physical inputs to the economy as a whole could be limited either through increasing fee or tax structures, or the (more inefficient) use of regulatory bans and limits. Note that this assumes a continuing capability of national states, or global society generally, to impose such boundary conditions on firms, even—or especially—large transnationals.

16.4.2 Firms and Technology

Industrial ecology analysis indicates that achieving sustainability will likely require not less technology, but the evolution of better—and probably more complex and information intensive—technology. Moreover, the rate of degradation of natural systems (loss of biodiversity, degradation of agricultural lands, mining of groundwater) argues that the evolution of environmentally appropriate technology must occur relatively rapidly, or social, cultural, and economic systems may not be able to adapt gracefully to an environmentally constrained world. It is also generally the case that in free-market economies, technology as a broad competency resides in corporations (there are exceptions, such as the national laboratory systems of many countries, but these are relatively rare, and tend to focus on specific missions, such as space exploration or military technologies). Taken together, the hypothesis that technological evolution is critical to sustainability and the fact that technology generally resides in corporations imply that private firms (or a similar type of agent) will be critical to any movement toward sustainability.

Obviously, however, firms can fulfill this function in many ways. They can be redefined so that movement toward sustainability becomes a specified internal goal (the mechanics of accomplishing this, especially as in many cases sustainability may conflict directly with at least the short-term profit motive, are unclear at best). They can be motivated by changing customer demand patterns to provide ever more environmentally appropriate products and services (this, of course, raises the difficult initial question of how to change customer preferences so that they result in behavior supporting a sustainable economy, an ideologically as well as practically difficult issue). Alternatively, firms can remain profit-oriented, and boundary conditions can be set around their behavior so that evolution of technology toward sustainability occurs.

In practice, a combination of these factors is coming into play. Some customers are, indeed, demanding environmentally appropriate products to the extent they can be defined (which is frequently far less than the public, environmentalists, or industry technologists realize). Some firms are exploring means by which uncosted externalities can be included in their management decisions, at least qualitatively. Governments are in fact establishing regulatory constraints, fee structures, and market information mechanisms (such as ecolabelling) which, intentionally or not, elicit environmentally preferable corporate behavior. These efforts, however, tend to be sporadic, internationally uncoordinated, substance specific, and localized in time and space; they do not yet reflect a sophisticated, coordinated effort to evolve technology, and firms, toward sustainability.

16.4.3   Firms and Social Costing

Increasingly firms are under pressure to include social costs, and not just economic costs, in their management decisions. As indicated elsewhere in this text, recent years have seen considerable progress in developing methodologies to determine the environmental impacts of substances and even products over their life cycles. Despite obvious problems in quantifying many of these impacts, the basic intent behind the development of such tools is that entities—including, prominently, private firms—use them for material, process, and technology choices, for product design, and for management decisions generally. Such tools would not be necessary if prices of inputs accurately reflected all externalities. The obvious implication, therefore, is that use of such methodologies will identify costs different from the economic costs upon which the firm traditionally relies or, in other words, capture externalities. The firm is then expected to modify its behavior based on the results of using the tool.

In one sense, this is simply seeking greater social efficiency within existing constraints. Thus, for example, if a DFE analysis establishes that one polymeric system is environmentally preferable to another in a particular application, and the costs are roughly equivalent, no economic penalty has been paid but social costs can be lowered. It is certainly true that, except in very rare situations, firms are precluded from choosing inputs, no matter how environmentally preferable, that make their products uncompetitive because of a concomitant cost, quality, or time-to-market penalty.

In another sense, however, this marks a fundamental questioning of the existing rationale of the private firm. Conceptually, the corporation is being asked to make decisions about its operations, products, and services based on something other than economic costs, including externalities that, although they may result from the firm's choices and operations, are not costed in the market. While it can be argued that decisions based on such criteria are, in fact, in the long-term interest of the firm—for example, because it adds to the value of the firm's trademark to be perceived as a good corporate citizen, or, more broadly, because it is in the firm's interest to ensure a stable path to a sustainable future, so that it may remain in existence and profitable—it cannot be denied that, at least in the short term, the firm is being asked to move away from a strict profit orientation.

16.4.4   Scale and Scope of Firms

The scope and scale of private firms is not coextensive with their environmental impacts. Virtually all modern approaches to environmental issues begin with the assumption that the appropriate scale of analysis is the life cycle of the material, product, or service at issue. Such a life-cycle approach is embedded in both the LCA and DFE approaches, and is clearly necessary and desirable if the inefficiencies and suboptimal results of localized optimization, at the expense of the performance of the overall system, are to be avoided.

On the other hand, there is an obvious disparity between the scope and scale of even the largest firms and the life cycles of their materials, technologies, products, and processes. Firms manufacturing complex articles do not typically extract or perform initial processing on the materials they use; that is done by petroleum, mining, or chemical companies. Manufacturing firms also do not usually manage the products after the consumer is through with them, nor do they manage the material streams that may result from the dismantling or recycling of post-consumer products. Conversely, firms providing raw materials for the economy seldom have a detailed idea of how their industrial customers are using, formulating, or disposing of excess material. Service firms usually do not understand the technology embedded in the products they use (telephone companies are not expert in designing switching systems, for example).

This structure of inter-related firms, each of relatively small scale, has arisen not only for economic and historical reasons—each firm seeking position in a relatively limited number of markets where its core competencies give it a competitive advantage—but also for legal reasons. In virtually all market economies, some form of antitrust regulation controls both the scale (size within markets) and scope (vertical combinations) of private firms, although the mechanisms and stringency of regulations vary considerably depending on the country. Moreover, in many cultures, especially that of the United States, large organizations in general are disfavored.

Environmental policies, however, are in the short term moving toward takeback policies for packaging and products, whereby the manufacturer is responsible for taking back its packaging and products after the consumer is through with them, and refurbishing, recycling, or properly disposing of them. Longer term policy discussions, especially in Europe, are beginning to suggest that manufacturers implement product life extension programs, in which products are refurbished and returned to commerce, as a step toward the "functionality economy," where firms sell functions, not products, to consumers. In such an economy, for example, an automobile company would lease cars to consumers, but remain responsible for the maintenance of the vehicle and all aspects of material management, from choice of inputs, to routine maintenance materials and lubricants, to recycling components and materials after the car is retired from service.

Whether it is through contractual arrangements creating a "virtual firm," directly as the primary customer interface, or as part of industry consortia establishing standards of performance and design, product takeback systems and the functionality economy would seem to imply a significantly expanded span of control for the firm in many instances. This need not be the case, however: In most developed countries, some 65 to 70 percent of automobiles by weight are recycled, and many subassemblies are refurbished and placed back into commerce without any explicit control mechanism. A wide variety of parts dealers, junkyards, scrap operators, and secondary smelters have formed a very effective recycling system that is organized only through mutual economic self-interest. An important area of study, therefore, is what conditions and technologies favor the evolution of such behavior. Why aren't white goods or electronic items similarly recycled? And what boundary conditions might result in the self-organization of a similar system for such products, or, alternatively, what policies might disrupt such systems and thus be disfavored?

Culture makes a big difference in the way industrial structure is viewed. In some societies, such as Japan, the existing economic structure, especially virtual integration of different firms in *keiretsu*, is structured so that antitrust issues and considerations of industrial concentration may not be a problem. In others, such as the U.S., it could be a substantial change from the status quo, and would likely face significant legal and political challenge. When combined with the newly globalized scope of commerce and the recognition that many environmental impacts are global in nature or implication, the pressure for larger commercial entities, which can be made responsible for their impacts broadly through time and space, is apparent.

16.4.5 Current Evolutionary Trends

In response to environmental trends, the behavior of private firms is already changing. Agents such as firms, after all, routinely adjust in response to changes in the conditions around them. The question is not whether change will occur; it is how fundamental that change will be.

The evolutionary changes that are already occurring will, however, make further change, even fundamental change, more easy. In this regard, it is interesting that the need for firms to integrate technology and environment throughout their operations, and recognize environmental issues as strategic for the firm, is arising just as the shift from regional and national to global economic competition, and from internal management based on mass manufacturing paradigms to information-based, non-hierarchical models of the firm, is occurring. Environmental issues then become one of a number of forces fostering radical change in the firm, and such change is easier when more than one driver is at work. It is therefore perhaps not surprising that industrial ecology is being accepted most rapidly in those industrial sectors (such as electronics) where other impacts of change have been particularly vigorous.

There are already a number of examples of corporate change in response to environmental concerns. Corporate codes of behavior, both at the firm level and at the trade group level, have proliferated over recent years: An early example is the

TABLE 16.1   Principles of the Business Charter for Sustainable Development

- Recognize environmental management as among the highest corporate priorities
- Integrate environmental policies and practices fully as a key element of management
- Continue to improve business's environmental performance
- Educate and motivate employees to carry out their activities in an environmentally sound way
- Assess environmental impacts before starting a new project or decommissioning an old facility
- Develop and provide products and services that do not harm the environment
- Conduct or support research on the impacts and ways to minimize the impacts of raw materials, products or processes, emissions, and wastes.
- Modify the manufacture, marketing, or use of products and services so as to prevent serious or irreversible environmental damage
- Encourage the adoption of these principles by contractors acting on behalf of a signatory company or organization
- Develop and maintain emergency preparedness plans in conjunction with emergency services and relevant state and local authorities
- Contribute to the transfer of environmentally sound technology and management methods
- Contribute to the development of public policy and government-business programs to enhance environmental awareness and protection
- Foster openness and dialogue with employees and the public regarding potential hazards and impacts of operations, including those of global or transboundry significance
- Measure environmental performance through regular environmental audits and relay appropriate information to the board of directors, shareholders, employees and authorities, and the public

Principles of the Business Charter for Sustainable Development, developed by the International Chamber of Commerce in 1991 (Table 16.1), and the CERES Principles developed by environmental NGOs and investor groups concerned about corporate environmental performance (Table 16.2). Another example is the Responsible Care Program and Product Stewardship Code developed by the Chemical Manufacturers Association (CMA) in the United States. Elements of these codes begin to reflect the trends discussed above. The CMA Product Stewardship Code, for example, includes a requirement that CMA members encourage distributors and direct product receivers to implement proper health, safety, and environmental practices, an indirect extension of the CMA member firm into the customer chain resulting directly from the desire to improve the life-cycle impacts of the product (in this case, at the use stage).

Other trade group initiatives have included the creation of guidebooks for environmentally preferable technologies for member companies. An early example of this activity was the Design for Environment Primer issues by the American Electronics Association in 1992, which consisted of 10 White Papers on various aspects of DFE (see Table 16.3). This was the beginning of a significant research effort aimed at institutionalizing life-cycle approaches to technologies, processes, products, and materials in the electronics sector. Thus, for example, the annual International Symposium on Electronics and the Environment program was begun in 1993 by the Institute of Electrical and Electronics Engineers, Inc., and the regular proceedings from those conferences are a valuable DFE resource to the practicing engineer.

TABLE 16.2   The CERES Principles for Corporate Environmental Disclosure

Introduction:

By adopting these Principles, we publicly affirm our belief that corporations have a responsibility for the environment and must conduct all aspects of their business as responsible stewards of the environment by operating in a manner that protects the Earth. We believe that corporations must not compromise the ability of future generations to sustain themselves.

We will update our practices constantly in light of advances in technology and new understandings in health and environmental science. In collaboration with CERES, we will promote a dynamic process to ensure that the Princples are interpreted in a way that accommodates changing technologies and environmental realities. We intend to make consistent, measureable progress in implementing these Princples and to apply them in all aspects of operations throughout the world.

1. Protection of the Biosphere

   We will reduce and make continual progress toward eliminating the release of any substance that may cause environmental damage to the air, water or the Earth or its inhabitants. We will safeguard all habitats affected by our operations and will protect open spaces and wilderness, while preserving biodiversity.

2. Sustainable Use of Natural Resources

   We will make sustainable use of renewable natural resources such as water, soils and forests. We will conserve nonrenewable natural resources through efficient use and careful planning.

3. Reduction and Disposal of Wastes

   We will reduce and where possible eliminate waste through source reduction and recycling. All waste will be handled and disposed of through safe and responsible methods.

4. Energy Conservation

   We will conserve energy and improve the energy efficiency of our internal operations and of the goods and services we sell. We will make every effort to use environmentally safe and sustainable energy sources.

5. Risk Reduction

   We will strive to minimize the environmental, health and safety risks to our employees and the communities in which we operate through safe technologies, facilities and operating procedures and by being prepared for emergencies.

6. Safe Products and Services

   We will reduce and where possible eliminate the use, manufacture or sale of products and services that cause environmental damage or health or safety hazards. We will inform our customers of the environmental impacts of our products and services and try to correct unsafe use.

7. Environmental Restoration

   We will promptly and responsibly correct conditions we have caused that endanger health, safety or the environment. To the extent feasible, we will redress injuries we have caused to persons or damage we have caused to the environment and will restore the environment.

8. Informing the Public

   We will inform in a timely manner everyone who may be affected by conditions caused by our company that might endanger health, safety or the environment. We will regularly seek advice and counsel through dialogue with persons in communities near our facilities. We will not take any action against employees for reporting dangerous incidents or conditions to management or appropriate authorities.

9. Management Commitment

   We will implement these principles and sustain a process that ensures that the Board of Directors and Chief Executive Officer are fully informed about pertinent environmental issues and are fully responsible for environmental policy. In selecting our Board of Directors, we will consider demonstrated environmental commitment as a factor.

10. Audits and Reports

    We will conduct an annual self-evaluation of our progress in implementing these Principles. We will support the timely creation of generally accepted environmental audit procedures. We will annually complete the CERES report, which will be made available to the public.

TABLE 16.3 American Electronics Association Primer: The Hows and Whys of Design for the Environment. This primer was one of the first source materials on DFE, and contained contributions from a number of leading firms.

American Electronics Association Primer: The Hows and Whys of Design for the Environment

	White Paper	Author
1.	"What is 'Design for Environment'?"	Brad Allenby, AT&T
2.	"DFE and Pollution Prevention"	Brad Allenby, AT&T
3.	"Design for Disassembly and Recyclability"	Robert G. Goessman, IBM
4.	"Design for Environmentally Sound Processing"	Janine C. Setukowski, AT&T
5.	"Design for Materials Recyclability"	Walt Rosenburg, COMPAQ and Betty Ryberg, Pitney Bowes
6.	"Cultural and Organizational Issues Related to DFE"	Brad Allenby, AT&T
7.	"Design for Maintainability"	E. Thomas Morehouse, USAF
8.	"Design for Environmentally Responsible Packaging"	Karen Rasmussen, General Electric
9.	"Design for Refurbishment"	Jack C. Azar, Xerox Corporation
10.	"Sustainable Development, Industrial Ecology and Design for Environment"	Brad Allenby, AT&T

Regulatory programs already in effect also reflect these trends. Germany has adopted a packaging takeback law, and recently concluded an agreement with the automobile industry to establish post-consumer takeback programs for automobiles sold after 1996; the Netherlands has also done so through an agreement, or Covenant, with relevant industrial associations. Post-consumer takeback of complex manufactured products is being considered or implemented in Germany, the Netherlands, Sweden, Austria, Denmark, France, Japan, and other countries, and is already a part of some voluntary eco-labeling schemes (such as the Blue Angel label requirements for personal computers).

Requirements in the United States that firms report their emissions of designated substances under so-called Community Right-to-Know requirements have had a strong effect on corporate behavior. Emissions of listed materials have declined substantially, and reduced emissions and pollution prevention are now part of the technology choice process for many facilities. Significant in themselves, such regulatory requirements, and their reflection in most of the industrial codes of behavior referenced above, are a first step in manufacturing becoming a collaborative effort among the firm(s) involved, its suppliers and customers, the community in which manufacturing occurs, and the host culture. Those who were not in manufacturing when each facility was a barony unto itself cannot realize what a fundamental culture change has already taken place within leading firms.

Finally, it is worth noting that another dimension of the private firm, its public relations apparatus, has also become far more active in this area than before. While this leads to the usual amount of "smoke and mirrors"—statements based less on technological advances and robust programs than on wishful thinking and the desire to appear socially responsible—it has desirable consequences as well. In particular, after a firm begins to view this as an area where it should appear progressive and takes public positions to that effect, it generates expectations not only on the part of its customers and external stakeholder groups, but also among its employees. The public relations positioning can thus become something of a self-fulfilling prophecy.

16.5   A PLANNED EVOLUTION OF THE PRIVATE FIRM?

As has been pointed out elsewhere in this text, industrial ecology often points to phenomenon that exist but have not traditionally been recognized by the policy apparatus, such as the increasing co-evolution of global human and natural systems. So it is here. The preceding discussion indicates that there are at least some powerful trends tending to drive a redefinition of the firm. These drivers encourage the creation of a firm that is larger in both scale and scope, and that has explicitly assumed at least some of the responsibility for mitigation of existing environmental perturbations and transitioning toward a more sustainable global economic state. The implications of such an evolution of the firm are substantial, and not all bad. In this section, some of them are reviewed, but as yet no obvious answer emerges.

To begin with, there is a difficult inconsistency in the demands an industrial ecology analysis indicates the future may well create. On the one hand, it is likely that the rapid evolution of environmentally appropriate technologies is necessary to avoid potential discontinuous shifts in natural systems (e.g., environmental catastrophes). Rapidly changing markets containing highly competitive firms tend to be the most innovative, and thus most supportive of such an evolution. On the other hand, as discussed above, the economic and social trends generated in response to those perturbations appear to imply the need for larger firms in more collaborative structures with significant, if unspecified, public interest components in their goal structures. Such an economic structure could well be less competitive, less innovative, and less conducive to rapid technology diffusion than today, particularly if public interest components are imposed by regulation rather than internally generated, and are thus liable to be rigid rather than flexible. Technologies will be frozen and incentives for innovation reduced, not expanded. Large, inflexibly regulated organizations are not noted for their innovative strength or capability to change rapidly.

Moreover, let it be assumed for discussions' sake that it is desirable to evolve towards firms that, in large part, have assumed the responsibility for achieving sustainability, and thus have developed the requisite power to perform the necessary functions, probably at the expense of the national state. The transition period would be quite difficult. Polls indicate that, relatively independent of country, the credibility of private firms on environmental issues is minimal at best, and public trust virtually nonexistent. How would firms be regulated over the transition period to assure that, on the one hand, they were able to meet accountability standards demanded by political reality, while, on the other hand, they were prepared over time to be legally, morally, and ethically responsible for the achievement of sustainability? This becomes a particularly important question given the devolution of power from national states to large firms and markets, because if governmental units cannot set general behavioral limits or boundary conditions on firms, at least during the transition period, then such an evolution is highly unlikely. Thus, the "sustainable firm" model may well require an international regime dominated by strong national states, which seems counter to at least some current trends.

On the other hand, there may be one significant advantage to the profit-driven model of the corporation: It is relatively easy for society to establish meaningful boundary conditions for such an agent. Simplistic economic models to the contrary, firms already incorporate many aspects of their cultures and societies within their operations, and are able to adapt more even as the dominance of the profit motivation is maintained. Moreover, maintaining the primacy of the profit motive in a sense maintains the natural selection pressures of the economy, which are arguably critical if rapid evolution within the system is sought. No externally imposed regulatory mandate can substitute for the constant pressure and brutal frankness of potential commercial failure. It may be desirable to maintain what the Austrian economist Joseph Schumpeter called the "gale of creative destruction" characteristic of capitalistic market economies precisely because of the need for creativity and evolution.

In the shorter term, it is also difficult at this point to understand the full effects of rapidly evolving private firm practices (e.g., agile manufacturing, just-in-time manufacturing, concurrent engineering, substitution of service offerings for product sales), as well as structure (e.g., virtual firms) on environmental performance. Changing economic conditions, such as the substitution of information for other inputs into the economy, and the increasing dominance of service sectors in developed economies, are also trends that may substantially change the environmental impacts of many firms. It is possible that continuing evolution of the private firm, combined with increasing understanding of the relationship between their behavior and perturbations of underlying natural systems, will lead to sustainable economic behavior without conscious intervention in the system (through, for example, deliberately changing the legal definition of a private firm).

If, however, private corporations are to remain narrowly defined, public policy must become far more sophisticated. What will be required is the establishment of boundary conditions that encourage the evolution, and especially diffusion, of environmentally appropriate technologies leading to the achievement of sustainability when those technologies cannot be defined until after the fact. Some aspects of such a policy can be relatively well defined: More and better data on emissions, environmental impacts, and their sources; a robust system of metrics tracked through local, regional, and global sensor systems; possibly properly designed product takeback and producer responsibility requirements; and price adjustments to internalize externalities, for example. In a broader sense, however, we need much more information and knowledge to do so. For example, it is arguable, if not apparent, that complex systems such as the economy tend to self-organize and exhibit distributed, not centralized, control, feedback, and internal regulatory systems. How can such distributed self-organizational behavior, deliberately targeted to the achievement of sustainability, be stimulated by appropriate policies? And how can policy be shifted to such a basis as a practical matter, when most existing environmental regulation assumes the opposite—the need for specific, mandated, central micromanagement of all behavior bearing on the environment?

The questions raised in this case study are both complex and somewhat conceptual, albeit of obvious importance. Both "sustainable firm" and "profit-driven firm" models raise interesting theoretical and practical difficulties in the context of a sustainable global economy. The next case study, which looks at the policy structure of the Netherlands, is thus a useful counterpoint indicating how one country has actually approached the question of sustainability in an integrated way, and how the public-private split of responsibilities and capabilities is seen in that context.

REFERENCES

AEA (American Electronics Association). *The Hows and Whys of Design for the Environment.* June 1993.

Allenby, B.R. "Industrial Ecology: The Materials Scientist in an Environmentally Constrained World." *MRS Bulletin* 1993, 17(3): 46–51.

Aspen Institute. "The Fading Influence of Government and National Boundaries as the Electric Utility Industry Restructures: How Far Will It Go?" (report of an Energy Policy Forum) Aspen, CO: 1995.

Costanza, R., ed. *Ecological Economics.* New York, Columbia University Press: 1991.

Daly, H.E. and J.B. Cobb, Jr. *For the Common Good.* Boston, Beacon Press: 1989.

Dertouzos, M.L., R.K. Lester, R.M. Solow, and the M.I.T. Commission on Industrial Productivity. *Made in America: Regaining the Productive Edge.* New York, Harper Perennial: 1989.

Drucker, P.F. "The Coming of the New Organization." *Harvard Business Review* January–February 1988, 45–53.

Friedman, M. *Capitalism and Freedom.* Chicago, University of Chicago Press: 1962.

McKie, J. W., ed. *Social Responsibility and the Business Predicament.* Washington, DC, The Brookings Institute: 1974.

OECD (Organization for Economic Co-Operation and Development). "Technology and Environment: Government Policy Options to Encourage Cleaner Production and Products in the 1990's," OCED/GD(92)127, Paris, 1992.

Rosenberg, N. and L.E. Birdzell, Jr. "Science, Technology and the Western Miracle." *Scientific American* 263(5): 42–54.

Rothenberg, J. "The Physical Environment." *Social Responsibility and the Business Predicament,* J. W. McKie, ed. Washington, DC, The Brookings Institution: 1974, pp. 191–216.

Saxenian, A. Regional Advantage: *Culture and Competition in Silicon Valley and Route 128.* Cambridge, MA, Harvard University Press: 1994.

Scherer, F.M. *Industrial Market Structure and Economic Performance.* Boston, Houghton Mifflin Company: 1980.

Stahel, W.R. "The Utilization-Focused Service Economy: Resource Efficiency and Product-Life Extension," in *The Greening of Industrial Ecosystems,* B.R. Allenby and D. Richards, eds. Washington, DC, National Academy Press: 1994.

EXERCISES

1. Write down your own definition of a private firm. Include at least five characteristics that differentiate private firms from government or academic organizations. Of the characteristics you have selected, what is the most important or defining one?

2. You are the Chief Environmental Regulator for Country Kendra, and have been given the job to minimize pollution in your country over a 10-year period. List and briefly

explain five policy mechanisms you would explore to accomplish this goal. Rank your five choices in order of:

    a.   The most effective in changing the behavior of private firms.

    b.   The most effective in encouraging the development and deployment of new, more environmentally and economically efficient technologies.

    c.   The most publicly acceptable.

    d.   The cheapest to implement.

3. Some have suggested that NGOs may become a critical component of international governance mechanisms. Based on existing patterns, what possible roles do you see environmental NGOs playing in a sustainable global economy? Which role do you think most likely? Will, or should, NGOs change if they are to assume this role? Defend your answer.

CHAPTER 17

Policy Case Study: The Netherlands

17.1 INTRODUCTION

Many of the issues dealt with in this text are complex and difficult to understand, and it is all too easy to conclude that progress in such an area will be impossible in practice. For this reason, a case study that serves as a proof of principle, and illustrates that reasonable policies and programs at a national level are both feasible and have been implemented, is worthwhile. The example chosen is that of the Netherlands, which arguably has the most intelligent policy system in the area of sustainable development and industrial ecology in the world. There are others—that of Germany and Sweden come to mind—but they are neither as far along nor as integrated as the Dutch model.

This is not to say that the Dutch system is perfect, nor that the conflicts among different stakeholder groups have all been resolved in the Netherlands. Indeed, the Dutch would be the first to admit that many problems, from the theoretical to the implementation of specifics, remain to be defined and addressed. It is also apparent that the Dutch approach cannot be directly adapted by a larger, more heterogeneous country such as the United States or Australia. Nonetheless, the scope, scale, and integrated nature of the policy system created under the Dutch National Environmental Policy Plans is both unique and indicative of the progress that can be made by serious, integrative policy initiatives looking toward sustainability.

17.2 POLICY OVERVIEW AND INDICATOR DEVELOPMENT

In 1989, the government of the Netherlands issued its first National Environmental Policy Plan, subtitled "To Choose or to Lose," based on the Dutch Bill on Environmental Policy Planning. As required by that legislation, and, not coincidentally, as an indication that the Dutch government had made the critical conceptual leap of regarding environment as strategic and not just overhead, the NEPP was signed not just by the Minister of Housing, Physical Planning and Environment, but by the Minister of

Economic Affairs, the Minister of Agriculture and Fisheries, and the Minister of Transport and Public Works. From the beginning, therefore, the integrative nature of the policy planning structure was apparent.

The NEPP specifically adopted sustainable development, as defined by the Brundtland Commission, as the premise for environmental management. The ideological issues that arise in other national states, such as the United States, regarding redistribution of income, population control, and other implications of sustainable development, are not as problematic in the Netherlands. Obviously, the government of the Netherlands, in adopting this goal, did not believe that a small country with airsheds and watersheds extending far beyond its borders and heavily linked into both European and global economic activity could be completely sustainable in a short period (one generation). If nothing else, for example, membership in the European Union, with its limits on environmental restrictions on trade, would make unilateral environmental standards difficult in many cases. Nonetheless, adopting sustainable development as a goal—much as it forms the basis of the industrial ecology intellectual structure outlined in Figure 2.1 in Chapter 2—both provided a foundation for their policy evolution, and encouraged them to ask the right questions.

Thus, the Dutch began by defining three different topic areas that required different approaches: issues, sectors, and regions. The issues were then further mapped into *themes*, or integrated issue frameworks, each of which had a major measurement unit called an *indicator* associated with it (Table 17.1). In turn, specific emissions were linked to each theme, which formed the physical connections between economic activity and environmental impact at a high level. In this way, definition of the problem in quantitative ways without sacrificing the systems-based approach could be enabled. As noted in Chapter 7, creation of this type of indicator system, which provides performance metrics linked to anthropogenic forcing of natural systems, is an important step in rationalizing environmental management at all social levels—consumer, firm, society.

It is important to note that some of these themes are probably broadly applicable to different countries, while others may be more idiosyncratic. For example, climate change and stratospheric ozone depletion are by their nature global problems, so the

TABLE 17.1   National Environmental Policy Plan Themes

Theme	Examples of Measurement unit	Emissions of concern
Climate change	Carbon dioxide equivalent (C eq)	CO_2, CH_4, N_2O
Stratospheric ozone layer depletion	Ozone depletion equivalent (O eq)	CFCs, halons
Acidification	Acidification equivalent (A eq)	SO_2, NO_x, NH_3
Eutrophication	Eutrophication equivalent (E eq)	Phosphates, nitrates
Dispersion of toxic substances	Dispersion equivalent (D eq)	Pesticides, heavy metals, radioactive materials
Solid waste	Waste equivalent (W eq)	Building, industrial, domestic, and agricultural waste
Disturbance of local environments	Nuisance equivalent (N eq)	Noise and odors

themes and their indicators are relatively universal. Eutrophication, however, is a critical problem for a country such as the Netherlands, which has high water tables throughout the country; many canals, streams, dams, and other water system components; and intensive agriculture. It would be a less important, perhaps even inapplicable, indicator for an arid country such as Morocco. Similarly, solid waste is an important indicator for a heavily populated, geographically small country such as the Netherlands, but less of an immediate problem in a country with abundant space such as the United States, Canada, or Australia.

17.3 TARGET GROUPS

Following the cascade from theory to actual practice, the identification of emissions linked to indicators and themes was tracked to major sectoral sources, called *targets*. The principle target groups are agriculture, traffic and transportation, industry, energy, refineries, building trades, and consumers and retail trade (Table 17.2). "Target" should not be taken pejoratively; the Dutch, as a trading nation, have a strong appreciation for

TABLE 17.2   Target Group Performance Indicators

Target group	Theme	Indicator
Agriculture	Acidification	ammonia emissions
	Eutrophication	phosphate emissions
	Toxic substances dispersion	pesticides
Traffic and transportation	Climate change	carbon dioxide emissions
	Acidification	nitrogen oxide emissions
	Disturbance of local environments	noise and odor pollution
Industry	Climate change	carbon dioxide emissions
	Acidification	sulfur dioxide emissions
	Solid waste	nitrogen oxide emissions
		industrial and chemical waste production
Energy sector	Climate change	carbon dioxide emissions
	Acidification	sulfur dioxide emissions
	Solid waste	nitrogen oxide emissions
		fly ash and slag production
		radioactive waste
Refineries	Climate change	carbon dioxide emissions
	Acidification	sulfur dioxide emissions
	Toxic substances dispersion	hazardous substances emissions
Building trade	Climate change	carbon dioxide emissions
	Toxic substances dispersion	CFC emissions
	Solid waste	creosote and harzardous substances emissions
		building and demolition waste
Consumer and retail trade	Climate change	carbon dioxide emissions
	Acidification	nitrogen dioxide emissions
	Solid waste	dumped household and white good waste

Based on A. Adriaanse, Environmental Policy Performance Indicators, Sdu Vitgeverij Koninginnegracht, 1993.

the efficiencies of market mechanisms (as well as their deficiencies). Thus, from the beginning in the initial National Environmental Policy Plan (188), the government recognized that "[t]he government cannot solve environmental problems on its own, let alone prevent new ones," and that private industry, properly bounded within a market structure, would be the source of environmentally preferable technology:

> Market competition provides the most fertile breeding ground for the creative process of technological innovation. Technology policy should recognize this fact and make the maximum use of instruments which accommodate the market approach. Technological development will be necessary if the objectives of environmental policy are to be met. Technology is also vital for industry to gain and/or maintain competitive advantage. In other words, technology is 'the clean engine of our economy.' (National Environmental Policy Plan 2, 28).

This passage is worth quoting at length because it is both illustrative of the sophistication of the Dutch approach—unquestionable environmental credentials, but clearly understanding the role of technology in alleviation of environmental perturbations—and because it is contrary to the usual image of Dutch environmental regulation in countries such as the United States. Generalizing somewhat, it is ironic that it is the U.S. which has an environmental regulatory system characterized by inflexibility and a strong, almost ideological aversion to technology and the market, and the Dutch who have a market and technology based, collaborative, environmental management approach.

The approach to target industry sectors as laid out in the National Environmental Policy Plan Plus (63) involved five steps:

1. The formulation of industry environmental objectives for the years 2000 and 2010, and 1995 where possible;
2. The creation of a "declaration of intent" to form the basis for negotiation with the target sector;
3. The simultaneous launch of an information campaign directed at involved firms explaining the environmental objectives to be achieved and providing technical support (such as information on clean technologies);
4. Creation of an implementation plan as a result of government/industry negotiations, frequently in the form of a "covenant"; and,
5. Implementation by firms of technological and other measures to reduce emissions in line with the agreed upon plan.

17.4 COVENANTS

A unique feature of the Dutch approach is the use of *covenants*, voluntary, enforceable agreements with industry that establish targets for industrial performance but, unlike regulations, can be relatively easily changed to reflect advances in technological or environmental information or practice. Negotiations, based on a previously prepared database that integrates sector data with emission reductions targets, government and international requirements, and other agreements, involve the government of the Netherlands,

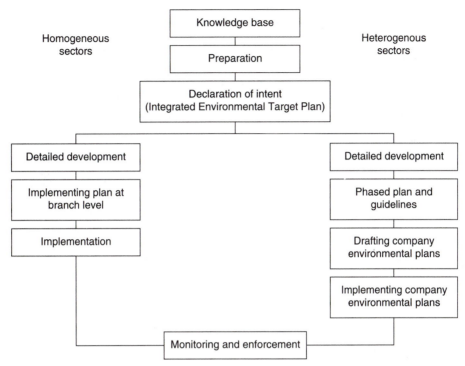

FIGURE 17.1 Covenant Process

provincial and municipal officials, and industry representatives, usually from trade groups or sectoral representatives. Sometimes trade unions are also involved. These negotiations lead to a Integrated Environmental Target Plan, which is then implemented by a process that differs depending on how complex the sector is (Figure 17.1).

Covenants are implemented through company environmental plans, which require the following, among other things:

- Best available technology as an assumed starting point.
- An inventory of emissions of concern and energy use.
- A summary of emissions reductions that have already occurred or will be realized under current legal requirements.
- A list of planned, additional measured reductions with a time table for implementation.
- A list of potential implementation problems, including financial, technological, and market-driven.
- Monitoring requirements.

These plans are public documents, and provide an important part of the transparency that is critical if such a flexible system is to succeed.

If another longish quote may be excused, it is interesting to note the value to both industry and the government that the covenant system is seen to provide:

Covenants offer industry greater certainty over a period of years, while avoiding the constraints of possibly inappropriate legislation. Companies know what they are required to achieve and by when; investment decisions and environmental policies can be planned with some confidence that the goalposts will not be moved. Within the target framework, business is largely free to determine its own priorities and balance sometimes conflicting needs. Covenants provide the opportunity for companies to influence measures and timetables during negotiations with government. Environmental targets applying to an entire sector minimize the risk of distorting competition within it.

Government is spared the need to prepare legislation which is difficult to draft in detail and expensive to enforce; the initiative for development (and monitoring) of environmental protection measures is largely transferred to industry. Industrial environmental objectives and published company environmental plans help the process of identifying the contribution individual companies should make to achieving the overall objectives of the NEPP. The reporting system associated with company plans enables government to monitor progress and to understand where problems are occurring. Not least, long-term agreements covering issues such as energy conservation and waste (traditionally hard to control under the permit system) will encourage companies to invest in more process-based measures to meet their environmental targets. (from Mattews, 11).

In short, the covenant system, and the collaboration and mutual understanding it generates, is a significant process step toward greater environmental and energy efficiency.

17.5 PRODUCT LIFE-CYCLE POLICY

It should not be thought, however, that the focus on emissions implied in the target sector system is the only dimension of the Dutch approach. Indeed, the government of the Netherlands, referring back to the touchstone vision, recognized in its Policy Document on Products and the Environment (PDPE) that "[t]he attainment of sustainable development requires changes in the patterns of consumption and production," not just production processes and manufacturing point sources, and that "[p]roducts directly or indirectly carry the environmental effects of the whole product chain." Accordingly, "[a]n integrated environmental policy aimed at modifying consumption and production patterns therefore needs to focus, amongst other things, on the product." Even here, the pragmatic approach embodied in Dutch policy is evident (7–8):

In terms both of its importance to trade and industry, and of its role in the life of the consumer, the product is part of the very fabric of Western culture. Much of the business community seeks to achieve continuity by providing the products demanded by the market; for consumers, many daily needs are satisfied by the purchase and use of products. Any improvement to the environmental quality of the product must not impair its ability to fulfill these functions. Environmental considerations are not generally the only factors on which either producers or consumers base their decisions. Aspects such as quality, functionality and price are also important. The aim of the policy outlined here is to ensure that environmental considerations achieve at least equal influence. Better still, the interests of the environment would be greatly served if it could be ensured that products of a higher environmental quality more closely met the requirements of the various parties.

Several obvious issues arise when the focus of environmental management systems, both governmental and private, shift from manufacturing emissions control to product orientation. The most obvious is the need for a life-cycle approach to both materials and products. This approach is captured in the "integrated substance chain management" approach for materials, and the "product-oriented environmental policy" approach for products. Both begin the process of defining methodologies by which such impacts can be identified and prioritized, in line with the themes and indicators already identified.

Another is the role for information. Adequate information is, of course, a prerequisite for economically efficient markets, especially where prices are known not to capture critical externalities. Accordingly, an important point of the product policy is the provision of adequate information on environmental impacts of products and constituent materials along the supply chain and to the final consumer. In keeping with the free market orientation, product information will "only be imposed [by statute] if and insofar as self-regulation proves to be inadequate." (PDPE, 12).

Another obvious dimension of the implementation of product life-cycle management is the need to consider the international dimension. There are three aspects to this. The first, obviously, is that the environmental perturbations of concern are not limited by national boundaries, and so must be addressed on the international scale. This is balanced by the difficulty of imposing Dutch standards on others: "the intention is not to pursue an extraterritorial environmental policy," (PDPE, 13). This is particularly true given the second aspect of product regulation: its impact on international trade. Global and European agreements limit the restrictions that national environmental regulation can put on free trade. The third aspect is the position of the Netherlands as a relatively small country highly dependent on international trade:

> An international approach is also economically advisable. The products sold in the Netherlands come from all over the world, and Dutch-made goods similarly find their way into markets the world over. National policy will therefore—partly in recognition of the need to preserve the international competitiveness of Dutch industry—take full account of the international dimension. (PDPE, 13)

Product regulation also inevitably raises the issue of material selection, and the concomitant question about how to evaluate both the embedded environmental impacts in materials, and the impacts of materials as they are used in specific applications. Methodologies such as "integrated substance chain management" were developed to understand such issues, but the goal of a database that contains such information in a high-level, useful format remains to be achieved.

17.6   PROGRAM SCOPE

It is not possible in a case study to cover all details of a complex government program, especially where the underlying substantive issues are so encompassing. A partial list of the reports that were planned in just the first several years on different aspects of the National Environmental Policy Plan, however, provides an indication of the scope and breadth of the activity engendered by this unique approach:

Risk Management
Tropical Rain Forests
Eutrophication
Radiation Standards
Organization Structure for
    Chemical Waste Disposal
Technology and Environment
Environmental Labeling
Transport of Hazardous Substances
Internal Environmental
    Management Action Programme
Collection/Processing of Used Oil
Manure Policy
Planning/Finance of Sewer Systems
Underwater Soil Issues
Long Term Research for
    Environmental Management
Plan for Synthetic, Natural, and
    Energy Intensive Materials
Soil Cleaning in State Waters
Target Group Policy and
    License Grants

Final Government Position on
    Brundtland Report
Acidification Abatement Plan
Ammonia Emissions Control
Multi-Year Crop Protection Plan
Drinking-Water Supply for
    Chemical Waste
Climate Change
Nature and Environmental Education
Water and Soil Quality
Dumping Policy
Prevention and Reuse of Waste
    Substances
Ten-Year Soil Cleanup Plan
Region-Oriented Environmental Policies
Sustainable Buildings
Environment and Eastern Europe
Products and the Environment
Energy Savings and Efficiency
Buildings and the Interior Environment
Instruments, Enforcement and
    Expansion of Environmental
    Protection

There are two important points to be drawn from this list. The first is the comprehensive nature of the activities, ranging from product regulation to trade to building codes and regulations. This program is, therefore, an illustration of what it means when environment is treated as strategic, rather than overhead, for society.

The second is the value of creating an intellectual structure within which the right questions can at least be asked. No one, certainly not the government of the Netherlands with their sophistication, would pretend they have answered all the relevant questions, or, less plausibly even, have achieved a rigorous definition of sustainability. Indeed, most aspects of this program are criticized by different stakeholders, from industry to NGOs. The use of sustainable development as a vision, though, and its expression in themes and quantitative indicators, has generated an approach to environmental issues which remains unique in its scope and sophistication.

17.7   ROLE OF TECHNOLOGY

In discussing the role of technology, it is important at the outset to differentiate between two very different kinds of technologies, which go under a confusing set of labels. As shown in Table 17.3, the first may be called "green technology," or "environ-

TABLE 17.3   Technology and Environment. "Green technology" is an ambiguous term that is sometimes intended to mean "sustainable technology" but in practice frequently reduces to compliance or remediation technology. Once environmental considerations are treated as strategic rather than overhead, however, they will become appropriate dimensions of all technological systems.

Technology Designation	Philosophy	Examples	Origin	Market Structure	End Point
"Green Technology"	"The government can mandate everything we need"	Scrubbers; water treatement plants; pollution prevention	Localized effects; command-and-control, end-of-pipe	Central mandate and control	Reduce local human risks
"Environmentally Preferable Technology Systems"	"Evolution of complex systems within appropriate boundry conditions"	Central digital servers providing video/music/multi-media on demand; energy and water efficient dishwasher	Technological change; Industrial Ecology theory and Design for Environment (DFE) implementation	Internalization of externalities; "free market" function within boundry conditions	Sustainability

mental technology," or, more accurately, "control technology." It is the familiar end-of-pipe emissions control and treatment technology, sometimes extended to in-plant easy process upgrades. Such technologies are a creature of, and are sustained by, a command-and-control approach to environmental regulation. In many countries, the lobby for this industry sector is one of the strongest in arguing for enhanced end-of-pipe environmental regulation.

Such technologies are obviously necessary in many cases, particularly in the short run before the underlying process and product technologies can be reengineered to reduce their environmental impacts. Nonetheless, they are profoundly inadequate for achieving increased economic and environmental efficiency, much less sustainability. They represent the continuation of the overhead paradigm, not the understanding of environmental issues as strategic for a firm or a country.

The Dutch have clearly recognized the differences between the two types of technology, and the inability of best current practices and incremental improvements to existing technology systems to achieve the improvements in environmental efficiency they anticipate will be required to achieve their goal of sustainability (recall Figure 11.1 from Chapter 11). Moreover, the models clearly recognize that the new suites of technology that will be required will not be available in the short term—less than 20 years—and that research on such technologies has just begun. Development and diffusion of such sustainable technologies remains in the future.

This evolution to new technology systems in the Dutch model incorporates elements of sustainable development such as regional distribution issues and increases in population and average prosperity. It involves the concept of "The Jump" (Figure 17.2), where incremental improvements in existing systems and industrial ecology research and development provide the basis for the necessary improvements in environmental capacity.

The planning method used to understand "sustainable technology" and the social, political, economic, and legal context within which such technologies must be evolved is called *backcasting* as opposed to forecasting. Rather than extrapolating forward, this method looks back toward the present from a future point—say, the year 2040—and

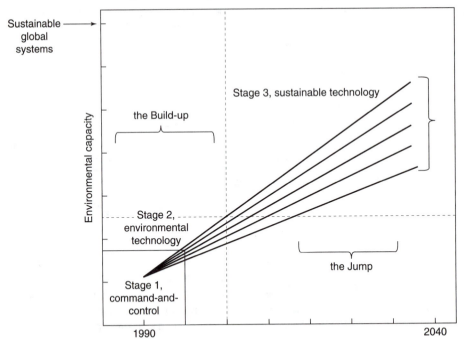

FIGURE 17.2 Stages for Realizing Sustainable Technologies
Source: Based on J.L.A. Jansen, "Technology for Sustainable Business in the 21st Century"
Presented at Globe 1994, Vancouver.

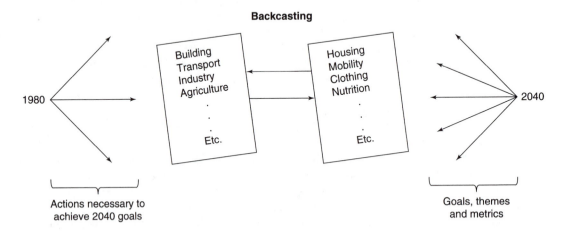

FIGURE 17.3 Backcasting. Backcasting is a planning methodology that asks where one must be at a future point (2040 in figure), then, given current conditions, looks backward to seek to identify paths and activities that can lead to that point.

asks what paths can lead to that point from the present (Figure 17.3). As explained by J.L.A. Jansen in *Sustainable Development: A Challenge to Technology!* (29),

Within the framework of technological development, "forecasting" concerns the extrapolation of developments towards the future and the exploration of achievements which can be realized through technology in the long term. Conversely, the reasoning behind "backcasting" is: on the basis of an interconnecting picture of demand which technology must meet in the future—"sustainability criteria"—the direction the process of technology development must take and possibly also the pace at which this development process must be put into effect can be determined.

Backcasting forms both an important aid in determining the direction technology development must take and in specifying the targets to be set for this purpose. As such, backcasting is an ideal search toward determining the nature and scope of the technological challenge which is posed by Sustainable Development, and it can thus serve to direct the search process toward new — sustainable — technology.

Interestingly enough, the backcasting planning methodology is now being considered by Sweden, another environmentally progressive country, as a potential tool to support its sustainable development programs and research activities.

17.8 PROGRAM EVALUATION

Any such comprehensive program will have contentious elements and receive mixed reviews from different stakeholders. Some environmentalists, such as Biekart (143), argue that the covenant system is too easy on industry, and his comments are a salutary reminder of the adversarial interests that no existing system has yet completely eliminated:

In reality, the present situation [of covenants and self-regulation] paints a different picture: there are regular street fights between environmental organizations, employers' organizations, companies and policymakers on these points. This commotion will probably continue for some time to come.

Nonetheless, it is also apparent that the policy structure erected by the government of the Netherlands is remarkably comprehensive and robust, and, perhaps most critically, actually links the sustainability goals of the far future to existing practices and requirements.

In this light, it is useful to briefly review the self-evaluation that the Dutch performed of their progress, as reported in National Environmental Policy Plan 2 (NEPP2). A number of successes were identified, including the manure and ammonia policy in the agricultural sector, the Packaging Covenant with the packaging industry, the construction of energy environmental action plans by energy distribution companies, and a number of energy efficiency agreements. While progress was satisfactory regarding construction and some agricultural issues, these areas were still regarded as somewhat uncertain, requiring long-term implementation agreements and customized implementation in many cases.

Perhaps more interesting, however, are the difficulties that the Dutch experience has revealed. For example, a consistent problem has been atomistic target groups: consumers, retailers, and small and medium size firms. Such groups have problems digesting complex information on life-cycle environmental performance, and their wide variety makes it difficult to target information to specific interests in a useful way. Moreover, achieving changes in behavior is more difficult for these groups.

Comprehensive sectoral initiatives have caused some difficulties. It is apparent, for example, that if groups are to be realistically expected to change their behavior, viable alternatives must be available. Using cars less frequently requires good public transit systems or bike path networks, for example. In other cases, the target groups might want to change their behavior, but market conditions have not yet generated adequate alternatives. For example, commercial hybrid vehicles—with base power provided by energy storage systems such as batteries or flywheels, and peak power provided by small fuel cells or internal combustion engines—do not yet exist even if consumers would prefer them to traditional internal combustion automobiles.

Certainty and predictability in policy formulation and implementation also proved to be important in practice, validating a claim many had made on intuitive grounds. In particular, confusion over priorities, uncertainties in future demands (and fear of ever-heavier ones), uncertainty over future costs of requirements defined in terms of increasing performance (e.g., reduced waste per unit), and failure to adequately coordinate policies among jurisdictions were identified as barriers to better performance.

In sum, NEPP2 concluded that:

> On the basis of experience with the implementation of the NEPP1, the prerequisites necessary for measures to be taken by target groups have become clearer:
>
> **a.** clear targets and sufficient information, so that target groups know how they are expected to modify their behavior, e.g. produce less waste.
>
> **b.** adequate technology and facilities. The target groups must always be in a position to modify their behavior. Thus, for example, a good public transport system is important if cars are to be used less.
>
> **c.** adequate degree of certainty: customized implementation. Target groups must know what is expected of them, and when. They also need to know which measures have priority. Finally it is important that central government should be asking for the same changes in behavior as provincial and local government. Better policy coordination is therefore important.

Perhaps the most important result of the experience of the Netherlands is not, however, the specifics of the program. Rather, it is the proof of principle; the demonstration that a reasonable, comprehensive program that integrates economic and environmental considerations across economic sectors is possible. Moreover, although there are of course questions of implementation, the usefulness of an appropriate conceptual framework, and the value of properly chosen metrics to knit together short-term performance requirements and longer term sustainability metrics, is also apparent. Such metrics can integrate short-term policy with longer term goals in ways that are both politically feasible and intellectually defensible. Finally, the reliance on covenants where possible provides the policy flexibility, which is so critical in a rapidly changing environment.

REFERENCES

Adriaanse, A. Environmental Policy Performance Indicators: A Study on the Development of Indicators for Environmental Policy in the Netherlands. Sdu Uitgeverij Koninginnegracht, 1993.

Biekart, J.W. "Environmental covenants between government and industry: a Dutch NGO's experience." Reciel 4(2):141–149, 1995.

Confederation of Netherlands Industry and Employers (VNO-NCW). Environmental Policy in the Netherlands: The Role of Industry. 1995.

Dutch Governmental Programme for Sustainable Technology Development. Looking Back from the Future. February 1994.

Heijungs, R., ed. Environmental Life Cycle Assessment of Products (2 volumes: Guide and Background). Agency for Energy and the Environment (NOVEM),1992.

Jansen, J.L.A., and Ph. J. Vergragt. "Sustainable Development: A Challenge to Technology." Ministry of Housing, Physical Planning and the Environment. VROM 92600/a/12–92, 1992.

Kuik, O. and H. Verbruggen, eds. In Search of Indicators of Sustainable Development. London, Kluwer Academic Publishers: 1991.

Mattews, E., compiler. "Environmental Policy in Action No. 1: Working With Industry." Ministry of Housing, Spatial Planning and the Environment. VROM 94066/b/3–94, 1994.

Ministry of Housing, Physical Planning and the Environment. "The Netherlands, National Environmental Policy Plan (To Choose or to Lose)." VROM 90312/6–89, 1989.

Ministry of Housing, Physical Planning and the Environment. "The Netherlands, National Environmental Policy Plan Plus." VROM 00278/10–90, 1990.

Ministry of Housing, Physical Planning and the Environment. "The Netherlands, Integrated Substance Chain Management." VROM 91387/b/4–92, 1992.

Ministry of Housing, Physical Planning and the Environment. "The Netherlands, National Environmental Policy Plan 2 (The Environment: Today's Touchstone)." VROM 94059/b/2–94, 1994.

Ministry of Housing, Physical Planning and the Environment. "The Netherlands, Policy Document on Products and the Environment." VROM 94196/h/5–94, 1994.

EXERCISES

1. Assume that one of the "themes" selected by your national state, which is physically similar to the United States, is "preservation of public land and associated resources for future generations."

   a. Develop a measurement unit (indicator) for this theme.

   b. Identify the components of the indicator for this theme. Justify your choices.

   c. What would be your "target sectors" to improve national performance regarding this theme?

   d. For what kinds of national states would such a theme be appropriate, and for what kinds might it be less useful? Consider both economic and physical dimensions in your answer.

2. Assume that an important policy of your national state is reducing emissions of chlorinated solvents, and that you, as a Government Assistant Minister of Environment, have been selected to negotiate a covenant with the dry cleaning industry to make such reductions (traditional dry cleaning methods result in significant emissions of the chlorinated solvent perchloroethylene).

   a. Based on the Dutch approach, what potential problems do you see arising as you begin your task?

    b.   Options to reduce emissions include i) using more modern, dry cleaning equipment with emission controls, ii) developing a new, water-based solvent system, and iii) redesigning fabric and clothes so that dry cleaning is no longer required. Which option promises the most robust control of emissions? Which option(s) will you most likely be able to negotiate with the industry? Why?

    c.   In general, cleaning clothes generates significant environmental impact whether dry cleaning or home cleaning is involved. Home cleaning, for example, uses substantial energy for hot water and air, and generates large amounts of waste water. Given this situation, should you expand your negotiations with this sector? How would you expect the suite of issues raised by cleaning clothes to be dealt with in such a negotiated covenant process?

3. You are the Minister of Technology for a progressive country which has adopted a goal of sustainability in one generation. You have begun to look at the issue of clothing. If the environmental impacts of clothes are considered comprehensively, a number of sectors—agriculture (for cotton and wool), transportation (for global distribution of clothing items), retail, and others—must be part of the evaluation. Based on the Dutch model, what policy structure would you put in place to assure a rational, and comprehensive, approach to this issue?

4. Using the backcasting technique, write a three-page essay on achieving energy sustainability in your local jurisdiction by the year 2020.

Environmental Security Case Study: The United States

18.1   INTRODUCTION

This case study is somewhat different from the others in that it explores in some detail a policy area that is not usually associated with industrial ecology: national security. This level of detail is provided for two reasons, the first being that the process by which two different policy structures—in this case, national security and environment—are integrated is an important dynamic that the industrial ecologist should understand. The second is that most of the industrial ecology methodology has been developed at the firm or sector level so far: Consider, for example, the Design for Environment matrix methodology of Chapter 15, and sectoral initiatives such as IPM (integrated pest management) in agriculture. An equally important dimension, however, is the understanding of industrial ecology research questions and development of appropriate data and methodologies at the national state level. Environmental security issues are one area where such methodologies are both necessary and are likely to be developed (see, for example, Figure 18.4), making this a case study of both methodological and substantive interest for the student of industrial ecology.

For many readers, the concept of "environmental security," or the integration of environmental issues and national security considerations at a national policy level, may well be novel, if not oxymoronic. It is, in fact, neither. Rather, it reflects the integration of two trends: that of environment becoming strategic at all levels of society, a subject with which this book deals, and the significant evolution of the national security concept in the post-Cold War era. Indeed, given that national security is one of the most strategic concerns of any national state, it can almost be said that environmental security is the penultimate recognition of environment as strategic. It is thus worth a somewhat detailed treatment.

This is not to say that the concept is well understood or, for that matter, even universally accepted as valid. It is a new enough concept that it is not in the mainstream for either the environmental or the national security community yet, and some of those who do know about it view such an integration with deep skepticism, even alarm.

In reviewing this policy integration, it is useful to remember that policy, including foreign policy, security policy, environmental policy, and science and technology policy, generally functions in the short term and focuses on the interests of a specific geographic area. Limits arise either from political structure, such as terms of office and the boundaries of national states—or, more fundamentally, from human psychological bounds (recall Figure 4.3). Most people don't think beyond a time horizon of a few years and a geographic range of miles, or, at best, their region. Many of the natural and human systems with which national security and environmental policy in the broadest sense must deal, however, lie beyond these intuitive boundaries. The evolution of successful national states, and, obviously, of many of the natural systems perturbed by human activity, occur over decades or even centuries. This case study, therefore, illustrates a key challenge that reappears throughout the text: The need to develop pragmatic policy systems that integrate gracefully and robustly over very disparate temporal and spatial scales.

This case study also illustrates the general pattern of adaptation as environmental concerns come into initial conflict with other social and legal structures that have previously failed to consider them. The first stage is usually initial conflict between two communities (in this case, the environmental and security communities), followed by negotiation and identification of the valid policy principles of both regimes, followed by creation of a new integrative structure. It is Hegelian in a way: The thesis of the existing legal structure (national security) is challenged by the antithesis of the newly recognized environmental requirements, which then combine in a synthesis that (in an ideal world) combines the appropriate elements of both. Thus, the concept of environmental security can be seen as one example, but not a unique example, of a dynamic occurring in many areas. It is a particularly interesting case study because it is still developing, and many of the specific issues have yet to be identified, much less resolved.

This case study will focus on environmental security issues from the perspective of the United States for three reasons. The first is that much of the initial policy development in this area has occurred in the United States, which thus provides a useful concrete example by which general concepts can be illustrated. The second reason is that environmental security, as a new policy area, is inherently ambiguous, and it is useful and reduces unnecessary confusion to use specific examples from a single perspective. Moreover, by definition security issues can only be identified with reference to the interests of a particular national state. The principles discussed in this case study in the context of the United States can be generalized, of course, to any national state.

18.2   THE MILSPEC/MILSTD OZONE DEPLETION EXAMPLE

An early example of this adaptation process will help make the broader and as yet nascent evolution of environmental security as a legitimate policy structure more concrete. The military in most countries is a large purchaser of goods and complicated weapons systems, and the manufacturing, design, and maintenance of these products is usually governed by complex sets of contracting, procurement, and operating requirements, including, in the U.S., Military Specifications (MILSPEC) and Military Standards (MILSTD). These requirements, of course, have been drawn up over the years to ensure appropriate performance of products and systems under the extreme conditions of military use, and have virtually never had any environmental inputs. They form a powerful and complex cultural and legal system. The interaction of this institutional structure with the efforts of the U.S. electronics industry to migrate away from use of chlorofluorocarbons (CFCs) in manufacturing products for the U.S. military is an interesting, if preliminary, illustration of the dynamics of environmental security.

Depletion of stratospheric ozone is, of course, a classic and elegant example of unanticipated impact of human economic activity on fundamental natural systems. In this case, anthropogenic gases, primarily CFCs, which are quite stable, were found to be migrating to the upper atmosphere where, subject to energetic sunlight, they released their chlorine. This element, in turn, catalyzed the destruction of stratospheric ozone. Ozone in appropriate concentrations in the stratosphere is important because it blocks highly energetic sunlight from hitting the earth's surface, where it can cause significant damage to living things. Once this relationship was understood, the international community accordingly crafted a response, the Montreal Protocol, which aimed to eliminate production and use of CFCs.

So far, so good. The MILSPEC/MILSTD regulatory structure protected the performance characteristics of military systems, and the Montreal Protocol responded to a serious environmental threat. But CFCs are not just an emission from certain industrial processes that can be controlled by a scrubber (thus treating environment as overhead). Rather, they were, at the time, a critical material in electronics and metal piecepart manufacturing—in other words, they were an integral part of the manufacturing complex. They were strategic to manufacturing, not overhead, and manufacturing of electronics items for the military was and is to a large extent governed by MILSPEC/MILSTD. And this created conflict between the two previously disparate regulatory structures of environment and MILSPEC/MILSTD.

Thus, it is perhaps not surprising that, when the American electronics industry began to phase out ozone-depleting substances pursuant to the Montreal Protocol, the single biggest barrier to prompt phaseout was not technological, not economic, not scientific: it was MILSPEC and MILSTD. In fact, because of cross referencing in government, industrial, and commercial documents, and use of the rigorously tested MILSPECs and MILSTDs as industry standards around the world, Tom Morehouse, then with the U.S. Air Force, estimated that half of all CFC-113 use worldwide for the manufacture of electronics circuit boards was driven by these standards. Moreover, it was not just manufacturing, but product design and maintenance during the product use phase that were an issue. Weapons systems like the C-130 aircraft, for example, had literally thousands of maintenance applications where the only acceptable process involved CFCs. A CFC ban thus directly affected military readiness and performance.

Overcoming this barrier did not imply ignoring the procurement system and the attendant specifications, nor reducing the technical rigor of performance requirements. After all, the policy rationale for this particular system—robust performance under adverse conditions—was both strong and continuing. Rather, the process involved the integration of environmental and performance requirements into a new generation of MILSPEC and MILSTD, which met the goals of both environmental and military procurement policy. This required a difficult and costly multiyear research program. This case study is, in effect, not just one of environment being recognized as strategic to the interests of society, but one of environmental security as well.

18.3 CHANGING DIMENSIONS OF NATIONAL SECURITY

With this as background, it is now useful to turn to that function which, for many national states, is the most critical: national security. The constellation of issues that support or threaten the fabric of a state and its territorial integrity are, virtually as a matter of definition, those of most concern. Two principle and comfortable assumptions that in the past decades supported the traditional view of such issues are (1) that the national state is relatively absolute, and, (2) since the beginning of the Cold War, that the conflict between capitalism and communism in various forms defined global geopolitics. These assumptions, at least in their absolute form, are becoming less valid. The termination of the Cold War, and its underlying bimodal global geopolitical structure based on rival nuclear superpowers with clearly opposed ideologies, has resulted in a more complex security environment. Regional and local historical, political, cultural, environmental, and economic pressures that were repressed during the preceding decades are now emerging, and, in conjunction with the loss of state control of weapons of mass destruction, proving to be significant sources of potential security threats and destabilization of existing states.

The degree to which these changes are viewed as real or lasting varies; there are significant differences in individual and institutional perception and the importance given to various potential trends and developments. Some, like Jessica Mathews, believe that the changes are fundamental, and that global civil society is being redefined:

> The end of the Cold War has brought no mere adjustment among states but a novel redistribution of power among states, markets, and civil society. National governments are not simply losing autonomy in a globalizing economy. They are sharing powers—including political, social and security roles at the core of sovereignty—with businesses, with international organizations, and with a multitude of citizens groups . . . . The steady concentration of power in the hands of states that began in 1648 with the Peace of Westphalia is over, at least for a while . . . .
>
> Increasingly, resources and threats that matter, including money, information, pollution, and popular culture, circulate and shape lives and economies with little regard for political boundaries. International standards of conduct are gradually beginning to override claims of national or regional singularity. Even the most powerful states find the marketplace and international public opinion compelling them more often to follow a particular course.
>
> The state's central task of assuring security is the least affected, but still not exempt. War will not disappear . . . [n]ontraditional threats, however, are rising—terrorism, organized crime, drug trafficking, ethnic conflict, and the combination of rapid population growth, environmental decline, and poverty that breeds economic stagnation, political instability, and, sometimes, state collapse . . . .

Under these circumstances, the Cold War operating definition of national security based on a bipolar world and primarily military confrontation is thought by many to be too limited. In particular, if environmental issues and perturbations are strategic to a society, one would expect them to become a prominent dimension of national policy.

18.3.1 Environmental Security as U.S. Policy

This is indeed happening, at least in the U.S. For example, in 1996 the Clinton Administration noted that:

The decisions we make today regarding military force structures typically influence our ability to respond to threats 20 to 30 years in the future. Similarly, our current decisions regarding the environment and natural resources will affect the magnitude of their security risks over at least a comparable period of time . . . . Even when making the most generous allowances for advances in science and technology, one cannot help but conclude that population growth and environmental pressures will feed into immense social unrest and make the world substantially more vulnerable to serious international frictions.

The Secretary of State at the time, Warren Christopher, reaffirmed this theme in a subsequent speech at Stanford University on April 9, 1996:

> . . . our Administration has recognized from the beginning that our ability to advance our global interests is inextricably linked to how we manage the Earth's natural resources. That is why we are determined to put environmental issues where they belong: in the mainstream of American foreign policy.
>
> . . . The environment has a profound impact on our national interests in two ways: First, environmental forces transcend borders and oceans to threaten directly the health, prosperity and jobs of American citizens. Second, addressing natural resource issues is frequently critical to achieving political and economic stability, and to pursuing our strategic goals around the world.
>
> In carrying out America's foreign policy, we will of course use our diplomacy backed by strong military forces to meet traditional and continuing threats to our security, as well as to meet new threats such as terrorism, weapons proliferation, drug trafficking and international crime. But we must also contend with the vast new danger posed to our national interests by damage to the environment and resulting global and regional instability . . . . A foreign policy that failed to address such [environmental] problems would be ignoring the needs of the American people.

Thus, it is fair to conclude that, at least for some leaders in the Clinton Administration, the need for an "environmental security" policy in some form was increasingly recognized and accepted.

It was less clear that the dimensions of this requirement, and the means by which it could be institutionalized in existing policy structures, had been adequately worked out. Many Clinton Administration statements assumed a very broad definition: the three goals of its national *security* strategy, for example, were:

> *Enhancing Our Security.* Taking account of the realities of the new international era with its array of new threats, a military capability appropriately sized and postured to meet the diverse needs of our strategy, including the ability, in concert with regional allies, to win two nearly simultaneous major regional conflicts. We will continue to pursue a combination of diplomatic, economic and defense efforts, including arms control agreements, to reduce the danger of nuclear, chemical, biological and conventional conflict and to promote stability.
>
> *Promoting Prosperity at Home.* A vigorous and integrated economic policy designed to put our own economic house in order, work toward free and open markets abroad and promote sustainable development.
>
> *Promoting Democracy.* A framework of democratic enlargement that increases our security by protecting, consolidating and enlarging the community of free

market democracies. Our efforts focus on strengthening democratic processes in key emerging democratic states . . . .

This definition, carried through in other Administration documents, encompasses a broad range of potential threats and issues, including but not limited to economic development, trade, and, included in the concept of sustainable development, virtually all regional or global environmental perturbations. Many in the traditional national security community (sub silentio for the most part) view these definitions as far too broad, even as they may accept them as legitimate foreign policy issues, at least on a case-by-case basis. After all, there are many resource scarcities and environmental perturbations around the world, most of which would impact the United States only minimally. Moreover, the resources to respond to challenges in these areas are limited, and their allocation must be prioritized to ensure that national security is not jeopardized, and that optimum benefit is obtained for their use.

18.3.2   Environmental Foreign Policy versus Environmental Security

This raises a critical point. In cases such as this, where two communities that are unfamiliar with each other, and two issue areas that have been previously disparate, are being integrated, achieving analytical clarity is very important to reduce unnecessary conflict (recall the continuing conflicts between the trade community and the environmental community over expansion of regional and global free trade agreements). Arguably, these initial efforts to integrate security and environment, while important policy statements, lacked such clarity. It is very important in this instance, for example, to differentiate between the perspectives of a global view, where one views human security or, more broadly, biological security as a whole, and a national state view, which focuses on the interest of the national state rather than global systems (see Figure 18.1).

Even at the level of the national state, it is necessary at a minimum to differentiate between *national security* issues and *foreign policy* issues, the former being a limited subset of the latter. Consider the example of a disease in an African state which limits the capability of the state to develop economically by reducing the ability of the population to work and imposing substantial health-care costs. This is an obvious humanitarian concern, and may generate appropriate relief efforts, particularly by the international NGO community, which tends to take the global rather than the national state view. From the U.S. perspective, such a condition may well be a foreign policy concern of the United States. It will probably not, however, be viewed as a national security issue.

On the other hand, consider a hypothetical destabilization of Chinese agricultural production as a result of changes in precipitation patterns, which could lead to augmented internal and external population migrations of some magnitude. Again taking a U.S. perspective, such a situation clearly raises humanitarian issues appropriately dealt with through foreign policy initiatives. Given the geopolitical positioning of China, and the economic importance of Asia as a critical element of the global economy to which the U.S. is tightly bound, this would also quite clearly be a potential national security issue as well.

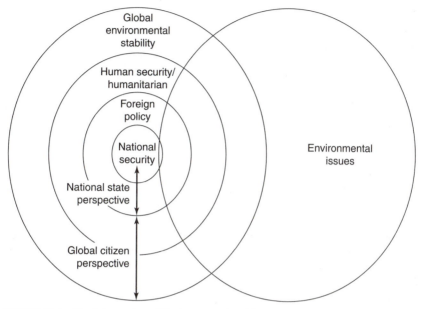

FIGURE 18.1   The Intersection of Environmental and Security Issues. While environmental issues cut across jurisdictional boundaries, the appropriate institutions for addressing them may not. In addressing such issues, therefore, it is important to understand whether the global citizen or national state perspective is appropriate. Even within the national state, issues may raise humanitarian or foreign policy considerations without raising national security concerns.

The nub of the problem is, of course, that there is no commonly accepted definition of national security. *Webster's New World Dictionary* defines "security" in relevant part as "protection or defense against attack, interference, espionage, etc. (funds for national *security*)," which, although seemingly specific, enables enormous leeway: "interference" is a subjective term, and the "etc." allows in what the rest of the definition might preclude.

Perhaps the most measured definition, which also captures the ambiguity of the term, is provided by Jack Goldstone, who notes in the Environmental Change and Security Project Report:

> There is only one meaningful definition of national security, and it is not inherently military, environmental, or anything else. Variations of that definition guided us throughout the cold war, and long before. That definition goes something like this: A 'national security' issue is any trend or event that (1) threatens the very survival of the nation; and/or (2) threatens to drastically reduce the welfare of the nation in a fashion that requires a centrally coordinated national mobilization of resources to mitigate or reverse. While this seems common sense, it is clear from this definition that not any threat or diminution of welfare constitutes a national security threat; what does constitute such a threat is a matter of perception, judgment, and degree—and in a democracy, a legitimate subject for national debate .... What has begun is an empirical assessment, within an existing and long-reasonable definition, of whether environmental trends, because of their threat to our survival or welfare, *must* be given attention according to this definition.

18.3.3   Collaborative versus Adversarial Approaches to Post Cold War Security Issues

It is important to realize that designating a set of issues as national "security" issues in no way implies the necessity of an adversarial approach, although this was usually the case during the Cold War. Rather, at least for OECD countries, it increasingly identifies areas where collaborative confidence building measures are increasingly being used. This broader, less adversarial positioning of national security, which increasingly is reflected in international and national agreements and collaborations, is in fact an important cultural change that greatly facilitates communication and cooperation between the environmental and security communities.

An important way to build such collaborative relationships is to focus on those dimensions of issues that can be seen as at least partially empirical, requiring both intellectual structure and data gathering and assessment to answer. This implies a necessary role for science and technology (S&T), and for research and development, a theme developed in Chapter 8. Developing such focused knowledge through appropriate research and development activities fulfills the critical need, given limited resources, to create a filter mechanism that can identify and prioritize potential environmental security issues. Common sense, for example, dictates the policy principle that, all things equal, investment in relevant research should primarily be directed at creating a targeted S&T base that defines and supports specific critical elements of an enhanced national security mission, rather than being scattered across all potential foreign policy issues, or even potential environmental security issues. This would appear to be a fruitful approach, at least initially. Rather than immediately jumping to the level of ideological confrontation, it reduces unnecessary conflict by first asking what issues can be resolved through empirical assessment and greater scientific and technological knowledge, and what issues properly remain in the domain of ideology and politics.

18.4   THE ROLE OF INSTITUTIONAL CULTURES AND CAPABILITIES

Industrial ecology is not only an integrative science, but an applied one as well. Accordingly, culture and institutional issues will frequently be important dimensions of industrial ecology investigations, as objective elements of the study in themselves. In this light, it is useful to consider some of the aspects of culture and institutional conflict and evolution that this case study illustrates.

The potential clash in underlying cultures between environmentalists and members of the national security community is apparent. Environmental NGOs often tend to be open, non-hierarchical, and liberal in ideology. They also tend to have the global, rather than the national state, perspective, as well as some aversion to technology and traditional military activities. They are often pacifist. Conversely, the national security community in most countries is conservative, insular, heavily focused on military threats and challenges, secretive, and powerful. It also tends to focus on short-term, obvious problems that can be resolved by the deployment of force. In this, it simply reflects the nature of its mission. Culturally, such security communities are among the least likely to embrace environmental considerations, and, when they do so, it is only in a mission-oriented context. More immediately, some in the security community believe that environmentalists and environmental scientists, facing cuts in their research funding, are urging "environmental security" as a means to obtain funding from security research programs, which in many countries remain relatively robust.

And yet, even given this inherently somewhat adversarial positioning, agreement on a wide variety of issues is not precluded. Three general principles seem to facilitate such agreement:

1. Parties must be sensitive not just to each other's stated positions, but to both explicit and implicit agendas. Moreover, the institutional capabilities and competencies of each is critical in determining specific courses of action.

2. Agreement is facilitated by differentiating issues based on whether they are susceptible to scientific and technological, as opposed to ideological, resolution. Even if they can be resolved by factual analysis, however, issues can be further differentiated based on the institutional competency of the parties.

3. Agreement is also facilitated if specific examples or projects, either existing or as future case studies, can be identified. Differences that are quite divisive in the abstract can often be reduced or eliminated when an appropriate factual example can be adduced.

These principles may be fruitfully applied to this case study. Even if an environmental perturbation may pose a significant threat to a nation, it may still not be a national security issue if it falls outside the competency and culture of the national security community and its component institutions.

For example, assume that anthropogenic global climate change is both real and can be shown to have such substantial negative impacts on the United States that it clearly meets usual operational definitions of national security threats. An argument can still be made that it is not a "national security" issue, at least in toto. This is because the scientific and technological research and development capabilities necessary to understand and respond to the phenomenon would reside broadly throughout the civilian research community, not within the traditional security organizations (the Department of Defense and the CIA, for example). Moreover, the scientific process most likely to result in rapid development and deployment of relevant knowledge would be the traditional one of open dialog and peer review, not the more secretive one which tends to characterize science and technology within the security community. A National Science Foundation, not a Department of Defense, would be institutionally and culturally better positioned to support such a program. This does not mean, of course, that the security establishment would not have some specific concerns (e.g., would any critical allies or areas of the world likely be destabilized by sea level rise), only that the issue, taken as a whole, is best not viewed as a "national security" issue.

Another example is stratospheric ozone depletion resulting from anthropogenic release of chlorofluorocarbons (CFCs) and other ozone depleting substances. In the absence of mitigation (which fortunately seems to be occurring), some estimates of potential impacts include upward of a million new cancer cases annually in the United States alone, with concomitant substantial mortality and economic loses. Many other significant human health and biological (agricultural) impacts are also possible. Such occurrences would obviously constitute a significant threat to the citizens of the United States, yet virtually no one has argued that ozone depletion should be handled as a "national security" issue. Indeed, trying to do so might well have derailed the broad research and technology deployment effort with which industries in many different sectors responded to the challenge of eliminating CFCs from their operations. At the same time, as the discussion of MILSPEC and MILSTD illustrates, there are dimensions of the ozone depletion issue that had significant operational impacts on military operations and weapon systems, and had to be addressed by and within the military and security communities.

18.5   ENVIRONMENTAL SECURITY TEST

Given the preceding discussion, a three part test may be used to determine whether an environmental issue or perturbation should be considered as an "environmental security" issue:

1. Are the potential impacts of the environmental perturbation in question substantial enough to be considered a national security threat?

2. Are the links between the environmental perturbation and the relevant impact(s) relatively certain and proximate? For example, one might argue that a collapse of the Mexican tuna fishery might encourage increased migration of unemployed tuna fishermen to the United States, which might cause political problems in California, which might generate social unrest in that state. The framework of the suggested problem is so speculative, however, and the links between the potential cause and the effect of concern are so vague and uncertain, that it is hard to argue that the state of the Mexican tuna fishery is an issue of national security for the United States.

3. Even if the environmental threat is substantial, certain, and proximate, is the national security apparatus institutionally and culturally the most capable of mounting an effective response? And, if so, to all or only to selected dimensions of the threat?

18.6   AN OPERATIONAL DEFINITION OF ENVIRONMENTAL SECURITY

A first step toward facilitating a reasoned, operational approach to the issue of environmental security is to generate a useful analytical framework. While there are obviously a number of possibilities, environmental security can be functionally described as an amalgam of four conceptually separate components: resource security, energy security, environmental security, and biological security (REEB). Although there is necessarily some overlap among these components, and between them and traditional security concerns, the conceptual separation is instructive (Figure 18.2).

1. *Resource security* involves two subcomponents: (1) local or regional competition for scarce resources, or (2) patterns of resource flows and use. Resource issues in

Components of the Environmental Security Mission

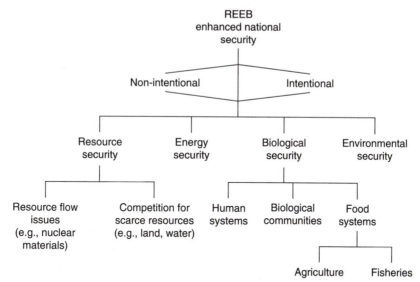

FIGURE 18.2   Components of the Environmental Security Mission. Such an analytical framework is useful in facilitating an operational, rather than ideological, approach to the enviornmental security concept.

either category become a resource security concern if they have the potential to give rise to political or military conflict involving the concerned national state. Competition for water resources or arable land are examples of the former, while a prime example of the latter is the management of flows of nuclear materials to avoid proliferation of weapons of mass destruction.

2. *Energy security* involves the identification and maintenance of access to energy sources necessary to support continuation of national economic and military activities. While military conflict deriving at least in part from competition over secure energy sources has already occurred (e.g., the Gulf War), public interest in energy security as an issue has waned because many assume (probably incorrectly) that energy markets are stable. Growth in demand, particularly in Asia, clearly threatens existing reasonable prices for, and access to, energy derived from various sources, particularly petroleum; conflict may arise because a stable, affordable flow of energy is critical to all developed economies.

Several points regarding energy security are worth noting. First, as with resources, absolute scarcity of potential energy resources is unlikely to be a concern. Rather, rapid fluctuations in supply and demand, local and regional scarcities, and the long lag times required to shift among different energy production and consumption technologies are the potential problem. Environmental and other social costs associated with energy production may also rise significantly as global energy markets expand substantially. Examples include greater frequency and amount of petroleum spills; increased leakage of natural gas from production, transportation, and storage facilities; and costs associated with management of nuclear power residual streams (the Yucca Mountain nuclear materials storage facility project in the United States has already cost some $1.7 billion).

3. *Environmental security* involves the maintenance of environmental systems, the disruption of which would likely create national security concerns for affected national states. Such issues could arise in either a domestic or foreign context. Examples might include releases of nuclear material in one state that over either the short or long term generate substantial impacts on other states, or environmental degradation in one locality that is so intense as to generate substantial population migration or other conditions with the potential to create conflict situations with neighboring states.

4. *Biological security* involves maintaining the health and stability of critical biological systems, the disruption of which would likely create national security concerns, either internal or external, for the affected state. The two most obvious classes of systems are (1) human populations and (2) food systems, including crops, livestock, and fisheries. A third, less obvious class of systems is biological communities of various kinds, such as wetlands, forests, or critical habitat, which frequently provide important "natural infrastructure" functions, such as flood control or fisheries breeding areas. A particularly difficult set of issues in this latter class arise when activity in one country affects an internal biological community, whose disruption has extraterritorial effects.

A potential biological security issue worth noting involves potential change in pathogen activity and distribution. Increased pathogen exposure and virulence due to changing cultural patterns (e.g., global travel), rapid evolution of bacterial resistance to antibiotics, and changing climate and human settlement patterns has been an increasing concern among experts. It is generally not realized how many previously unidentified infectious agents are still being detected. Since 1982, for example, 11 human diseases have been newly identified, including human immunodeficiency virus, hepatitis E virus, hepatitis C virus, Venezuelan hemorrhagic fever, Brazilian hemorrhagic fever, human herpesvirus 8 (Keposi's sarcoma), and HTLV-II virus (hairy cell leukemia). The possibility of significant domestic impact on human or biological system health as a result of new pathogen activity is one that, although downplayed, cannot be ignored.

18.7 INTENTIONAL AND UNINTENTIONAL PERTURBATIONS

A second classification of potential REEB perturbations, differentiating between *intentional* and *non-intentional* activities, is useful in constructing an enhanced national security structure. Both intentional and non-intentional activities must be considered as part of environmental security, but there is an important distinction. Non-intentional activities may or may not rise to the level of security issues, and, for institutional reasons if nothing else, may not lend themselves to a security community response. Intentional ones, on the other hand, frequently will, as they are, by definition, extensions of another state's policies and interests through deliberately chosen actions. Moreover, they are more likely to be difficult to counter, and more likely to constitute a significant threat as they presumably have been chosen to be effective.

The capability to generate such threats is augmented by the characteristics of the natural systems upon which they are based: Local actions can frequently perturb regional or global systems, thus permitting a state to project international threats based on internal activities. The Chernobyl incident is an unintentional example that involved only one facility but affected much of Europe. An intentional example is the concern during the recent Gulf War that the Iraqis would deliberately burn so much oil that the resultant particulates would lower global temperatures and sunlight penetration, creating a widespread ecological disaster. Rapid assessment by the U.S. Department of Energy national security laboratories indicated that such a threat was groundless, but, had such an S&T assessment capability not been available, the impact on planned military initiatives in the area might have been substantial. Moreover, even though the larger threat was groundless, the environmental conditions that were created during the conflict by the intentional burning of petroleum by Iraqi forces generated difficult military and personnel conditions, and the possibility that the health of U.S. and allied troops was affected is still under investigation.

Moreover, because of the technically complex nature of such systems, a relatively simple perturbation can have numerous and complex potential effects that are virtually impossible to counter once the system is perturbed.

18.8 STRUCTURING THE ENVIRONMENTAL SECURITY MISSION

An important part of any case study involving the integration of environment with other policy structures is the means by which the cultures and institutions can evolve gracefully toward the integrated system. In this case, it requires not only understanding the environmental structure, but the national security one as well. It is indicative of the importance of research, and S&T generally, that this forms an important mechanism for both supporting integration, and carrying out the function once integration is achieved.

The Cold War national security policy structure consisted of two closely linked primary components: an S&T base, which provided military capability, threat definition, and technological support for collaborative threat reduction (e.g., monitoring treaty compliance); and a policy component supported by that base. The structure required to support an environmental security mission is analogous, but perhaps not as widely recognized. In particular, it is necessary to build an S&T capability to support the development of the resource, energy, biological, and environmental components of national security.

Figure 18.3, based on industrial ecology principles, illustrates one possible policy/S&T framework for an environmental security mission. The first step in creating the S&T base for a particular issue is to understand the dynamics of the underlying physical system, which might include, for example, generating a model of its behavior. Depending on the system, such models may be fairly simple. On the other hand, they might be quite complex, as where, for example, an attempt to understand potential future precipitation patterns and water management systems in Asia would be part of a confidence building program with the goal of ensuring that crop failures and food shortages did not result in destabilizing population migrations. Such a model—more accurately, a nested set of models at the different relevant scales—might have to link

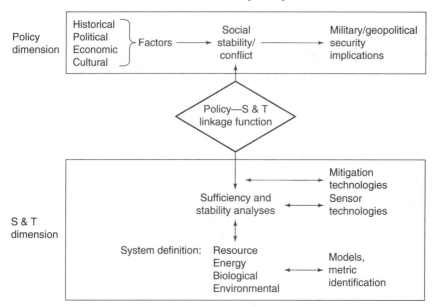

FIGURE 18.3   Environmental Security Policy Structure. It is especially important to make sure that the coupling between the science and technology dimension and the policy dimension is robust.

together a number of sub-models covering a wide spatial and temporal scale. An example of such a model set, which is essentially an industrial ecology methodology at the national state level, is provided in Figure 18.4.

As understanding of the system is gained, metrics by which one can evaluate its evolution over time can be developed. Ideally, such metrics will support the capability to predict when the system might be approaching instability, a particularly important concern because patterns of human activity tend to be predicated on the assumed stability of underlying natural systems, and much human effort is essentially aimed at engineering such stability into inherently variable systems. Thus, for example, much of the manipulation of rivers in the Middle East is intended to stabilize their annual supply of water at the highest possible level (these riverine systems are by nature highly unstable), as well as to expropriate as much of the resource as possible. Instabilities in natural systems, such as precipitation patterns, changes in groundwater flow, or other perturbations which either increase the interannual variability or reduce the amount of water which can be reliably produced, can under the circumstances generate the potential for resource scarcity conflict.

Once the system is defined and its behavior and stability assessed, sensor technology to provide input and track system evolution against the appropriate metrics can be deployed. Depending on the parameter, sensor systems may be either ground- or satellite-based.

The final step is development of mitigation technologies, including, if appropriate, traditional remediation technologies, before the potential conflict develops. In fact,

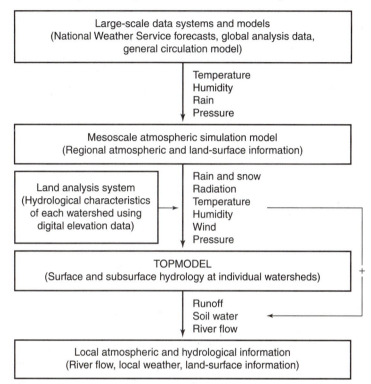

**Lawrence Livermore National Laboratory
Coupled Atmosphere-Riverflow Simulation System**

FIGURE 18.4   Lawrence Livermore National Laboratory Coupled Atmosphere-Riverflow Simulation System. Developing coupled model sets, particularly over such large spatial and temporal scales, is quite challenging and requires substantial computational power.

with luck the issue can be identified, defined, and resolved within the context of a collaborative S&T effort without rising to the policy dimension at all. For example, if crop failure resulting from changes in precipitation patterns is a concern, an entire set of mitigation efforts is possible, depending on the time scale. With several years warning, new crops and cultivars can be introduced that are more robust under the predicted conditions. Even with less warning, water recycling, demand reduction, and water storage technologies can be deployed. At the least, appropriate food transportation, storage, and distribution facilities can be prepared. Additionally, of course, a number of mitigating policies can be adopted by the international community based on the projected perturbation, including, for example, more planting of grain in other exporting countries to buffer the anticipated demand. Developing such systematic responses over varying spatial and temporal scales is an obvious application of industrial ecology, drawing on its emphasis on comprehensive, systems-based approaches.

Once the S&T dimension of a particular issue or set of issues is established, it is then possible to integrate the results into a robust security analysis and policy. While it is possible that a natural system perturbation, in itself, could generate national security implications, it is more likely that in many cases the national security effects of perturbations will arise only when they occur in conjunction with more traditional indicia of state instability, which themselves reflect specific historic, political, economic, and cultural factors. Thus, the S&T base does not replace, but is a necessary component of, enhanced national security policy considerations and analyses.

Experience indicates that the linkage between the S&T and policy dimensions, while conceptually apparent, is frequently weak or less effectual than possible. This might be particularly difficult as the organization providing the S&T capability will in most cases not be the organization making the policy decisions. It is therefore worth emphasizing the need to establish a clear linkage function between the S&T and policy dimensions at the outset.

18.9   PRIORITIZING ENVIRONMENTAL SECURITY ISSUES

The most important initial focus of any environmental security mission should be on existing or foreseeable intentional threats, and on those non-intentional issues that have already given rise or contributed to national security concerns. Again taking the U.S. perspective, one example is the well-known issue of water quality and water reallocation in the Middle East peace process. Another is North Korea, where nuclear material stocks and flows are a significant proliferation concern, and food shortages resulting from unusual precipitation patterns and flooding may be creating destabilizing conditions that, given the posture of the state, may lead directly to initiation of military conflict.

Such conditions, which almost by definition are giving rise to current national security concerns, must be addressed on an immediate basis. It is possible to structure a framework, however, within which enhanced understanding of the underlying physical systems, and perhaps technology development and deployment efforts, can be developed as a part of existing policy initiatives to reduce tensions and avoid escalation of conflict. Response to the North Korean situation, for example, has already included transfer of energy production technology designed to increase that state's energy security. Response to the food shortage issue might include not just the immediate response (provide food) but development and deployment of a more sophisticated conflict avoidance S&T strategy, to include developing models and sensor systems (probably satellite-based, under the circumstances) that can help predict when perturbations in underlying physical systems could impact food production and distribution.

From an industrial ecology and social perspective, the real advantage of the REEB approach, of course, is in its ability to reduce the possibility, severity, and expense of future conflict and environmental impacts. If this promise is to be achieved, a prioritized approach to those regions and issues that, at least initially, appear to offer the greatest potential impacts on the security of the relevant national state should be developed.

Using the United States as an example, it is possible to integrate several prioritization mechanisms in a simple format. It is apparent, for example, that some regions are more critical to U.S. national security than others. Moreover, some issues will be of more importance to the United States than others: Nuclear material flows, for example, will be a consistent resource concern globally. Finally, traditional indicia of environmental impacts—including the duration, severity, and geographical scope of the insult, and the technical difficulty and expense of mitigation—will also be important in prioritizing environmental security issues.

Application of these prioritization mechanisms to the set of potential issues cannot be done rigorously *a priori*. It is possible, however, to construct a matrix using these guidelines (Figure 18.5) that links five geographic areas of self-evident critical geopolitical interest to the United States—China, Mexico, the Former Soviet Union, Southeast Asia (including India and Pakistan), and the Middle East—with the four REEB categories. Where applicable, within each cell examples of issues that appear to be the most pressing are identified. Similar matrices could easily be generated by other countries as well.

While this structure should not be interpreted to imply that other geographical areas, or REEB issues, are not of concern, significant instability in any of these regions could have immediate and serious foreign policy implications for the United States. The mechanisms by which REEB forcing functions might impact states may vary—population migration, increased state instability—and the effects on the United States could be either direct (e.g., increase in NAFTA population migration, or diversion of

Environmental Security Prioritization Matrix

Region	Resource security	Energy security	Environmental security	Biological security
China	• Nuclear material • Commodity consumption patterns • Water	• Petroleum demand • Petroleum supply • Nuclear energy systems	• Environmental costs of economic growth	• Crop stability and food demand growth • Population stability
Mexico	• Water • Land distribution		• Environmental costs of economic growth	• Crop stability • Pathogen systems • Population stability
FSU	• Nuclear materials	• Nuclear energy production technology	• Environmental costs of economic growth • Nuclear waste issues	• Population stability
South-East Asia	• Nuclear materials • Water	• Petroleum demand • Petroleum supply • Nuclear energy systems		• Crop stability and food demand growth
Middle East	• Nuclear materials • Water	• Petroleum supply		

FIGURE 18.5   Environmental Security Prioritization Matrix. This illustrative matrix identifies environmental security priorities from the perspective of the United States. Similar matrices, with different regions and issues of concern, can be constructed for other national states. The resource, energy, environmental, and biological security categories would remain the same.

nuclear material to terrorist organizations) or indirect (e.g., instability in Asia or China causes regional economic dislocation, which in turn generates recession or depression in the United States). Nonetheless, the potential impacts of these particular issues on the United States and its citizens are, by-and-large, both apparent and potentially significant. To illustrate this point more specifically, two initial case studies, again from the perspective of the United States, can be suggested.

18.9.1   Water and Food in Mexico

Global population migration as a result of food shortages is a continuous and probably inevitable phenomenon that, in most cases, will not raise national security issues for the United States, although it may call for humanitarian foreign policy responses. There are a relatively few cases, however, where such migrations may have such direct impacts on the United States as to give rise to legitimate national security concerns. Substantial crop failures over a several year period in Mexico, induced by changes in precipitation patterns, may be a hypothetical example.

Mexico is currently undergoing rapid economic and political evolution as it adjusts to the accelerated regionalization of its economy, partially as a result of the North American Free Trade Agreement, and concomitant political evolution away from the paternalistic one party system that has characterized its governmental structure since World War II. Peasant technologies little changed for centuries, especially in the agricultural sector, coexist with modern industrialized facilities owned by transnationals competing in global markets. Cultural and legal systems that embed traditional class structures and support land-owning elites are increasingly challenged by modernist reformers, a conflict which in Chiapas led to armed confrontation between the Zapatistas and the state. Under these already somewhat unstable circumstances, crop failure as a result of water quality and/or quantity limitations may be a trigger for substantially increased internal unrest and consequent migration.

Two phenomena with national security implications result. Most obviously, a substantial increase in migration, both internal and between the United States and Mexico, could well occur. Since even existing levels of migration have already led to political conflict within some affected U.S. jurisdictions, this could cause serious social tension in the U.S. Predictable effects might include increased xenophobia and increased social tension, especially in border areas in California, and efforts to impose restrictions on migrants that have the effect of encouraging discrimination against American citizens of Hispanic descent.

Second, NAFTA is both a continuation and a recognition of a trend toward a regionally integrated economy including Canada, the United States, and Mexico (and perhaps others such as Chile). Disruption of these growing economic relationships would be costly both politically and economically, and would have the potential to generate a negative feedback loop: Increased economic hardship in Mexico would lead to increased migration pressure, which, in turn, would exacerbate the political and economic disruption of existing NAFTA arrangements.

In both cases, an important forcing function for destabilization in an already difficult situation (e.g., a weakened state in a period of economic and political transition) and consequent migration would appear to be perturbations to available water resources. Policies that more rigorously define, and concomitantly provide the basis for reducing, that forcing function are therefore desirable, all else equal. As in Figure 18.3, these policies will fall into two dimensions: the familiar policy dimension and the less familiar S&T dimension.

It is this S&T dimension which leads to the need for developing industrial ecology methodologies appropriate for environmental security analyses. In this case, the specific question can be asked: What S&T base should be developed to help support the stability of existing population patterns in Mexico and the border areas of the United States? Answering that question requires a systematic approach informed by the principles of industrial ecology (Figure 18.6).

1. A key driver for population migration is agricultural failure arising from patterns of distribution of two key resources—water and land—given existing populations and expectations. The link between the S&T dimension and the policy dimension thus flows through these categories. Both dimensions must be understood if the environmental security concerns are to be mitigated.

2. The S&T research program begins with development of a set of models that can be used to identify geographical and technological areas of greatest concern (e.g., where are resource conditions most marginal to begin with, and is there a crop or set of crops that are least stable under prevailing conditions). Such a system might begin by looking at existing precipitation patterns at a relatively high level with a global or, more likely, mesoscale model. Then, a set of

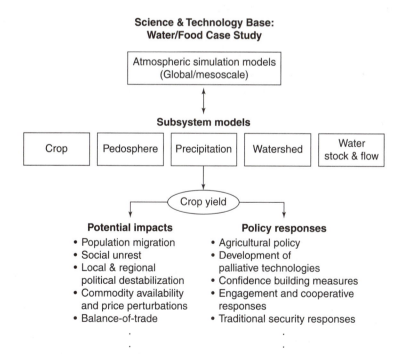

FIGURE 18.6   Science & Technology Base: Water/Food Case Study

subsystem models of crop distribution and response, soil systems (the pedos-phere), localized precipitation patterns, runoff and watershed response, and groundwater systems would be used to link precipitation with ability to support current agricultural systems, and determine whether, and to what extent, instability in precipitation patterns could generate meaningful agricultural disruption (i.e., disruption that would be significant enough to generate substantial pressures for migration or other potential impacts). Throughout this process, uncertainties of all kinds that might impact the prediction should be identified.

3. Once the baseline systems, including modeling and data components, are in place, a predictive capability would be built. The essence of this activity is to identify potential instabilities in resource availability before they occur, determine whether they would be meaningful if they occurred, and identify uncertainties associated with the prediction (some of which, like poor data, might be reduced by further research, while some, such as chaotic behavior of natural systems, might be irreducible within certain boundaries). For example, is it possible to tell from an integrated assessment of data and models when Mexican precipitation patterns in key areas are becoming unstable in such a way as to impact critical agricultural activities before the fact?

4. Concomitant with S&T system development and deployment is the need to deploy the appropriate sensor systems to monitor the physical system's state and performance. One might, for example, need data not just on precipitation and surface water flows, but on soil moisture content, vegetation stress, nutrient availability, and other parameters. Such sensor systems would probably include a satellite-based component, but, as always, ground-based verification of satellite data, and independent data generation regarding parameters that cannot be determined by remote sensing technologies, would be desirable.

5. The final step would be collaborative programs to develop and deploy mitigating technologies, which could range from engineering new varieties of existing crops, to introduction of new species entirely, to water or resource recycling or replenishment technologies. This step in particular must be linked to policy and state initiatives, as technology is a cultural as well as an engineering phenomenon, and inappropriate technologies are unlikely to be successfully deployed.

18.9.2 Nuclear Materials

The example of nuclear materials provides an illustration of the use of the traditional industrial ecology methodologies known as *industrial metabolism studies* in an environmental security context. Nuclear materials are an example of a resource security issue that cuts across both traditional national security and environmental interests, and includes significant energy security and biological security dimensions. Their inherent characteristics, uses, management, and impacts as improperly handled waste raise some of the most difficult and complex issues in the modern world. They are the basis for nuclear weapons, which previously were reserved to a limited number of states, but are now potentially available to terrorist organizations. They also are the basis for nuclear power, a technology that almost certainly will be increasingly deployed in the future, especially in Asia where economic expansion is driving an almost desperate increase in demand for energy production. The science and technology surrounding them in virtually any application is complex and arcane, as shown in Figure 18.7, which provides a high-level overview of the nuclear fuel cycle, while the politics are polarized and bitter.

Nuclear waste, whether in the United States, the FSU (Former Soviet Union), or, increasingly, in Asia, potentially poses some of the most serious real risks associated with environmental pollution, and cleanups are both expensive and technically challenging, where they can be done at all. Some contamination incidents—Chernobyl being the classic case—have caused extensive regional contamination, which, had it not obviously been unintentional, might in itself have been a trigger for conflict. The impacts were enormous: Within the Ukraine alone, 135,000 people were displaced within 10 days as a direct result of the incident, a figure that has since grown, and over 5 percent of that state's area remains significantly contaminated. The continuing destabilizing effects of that incident are demonstrated by the fact that, even now, the Ukrainian government, in a severe economic crisis, must continue to spend more than 5 percent of its budget in dealing with the continuing impacts of Chernobyl, including, for example, providing emergency housing to over 3 million directly affected people in Ukraine alone.

Dealing with nuclear materials issues is difficult in part because of their military (and terrorism) implications. Civilian stocks and flows of such materials are linked inevitably with military and security concerns and the potential for "environmental terrorism": A terrorist group would not need to explode a weapon, but could simply distribute radioactive materials widely in a heavily populated area, to achieve an impact.

Of course, a number of scientific, technical, and political efforts have been made to reduce the risks that nuclear materials pose in various military and civilian applications, and a number of national and international organizations, including the International Atomic Energy Agency (IAEA), are active in supporting that goal as well. Nonetheless, the effectiveness of such efforts is limited by activities of rogue states (e.g., North Korea, Iraq) and a lack of resources (e.g., to provide alternative energy production sources to replace allegedly unsafe reactors, or support IAEA activities at a sufficiently high level).

Given this situation, a possible environmental security policy for nuclear materials can be developed based on the obvious recognition that safe global management of such materials is an important component of U.S. national security. An important focus would be on the development and deployment of an S&T strategy that would both reduce risks, and, in many cases, provide an important vehicle for developing collaborative and confidence building exercises with other states (an important goal given that many of these states are either actually or potentially nuclear powers). Such a program would consist of several major components.

1. Construction and maintenance of a global database and model system capturing the stocks and flows of as much nuclear material as possible. Such a system should be driven by a need to understand the physical structure of the "industrial metabolism" of these materials, not by, for example, relatively arbitrary regulatory distinctions between different kinds of "wastes" or regulatory regimes. It

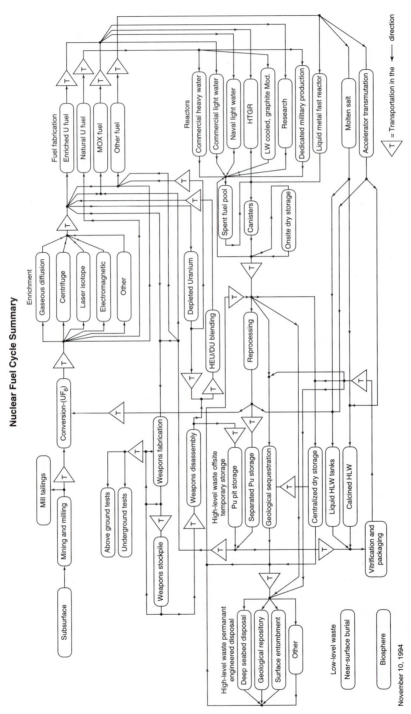

FIGURE 18.7   The nuclear fuel cycle is quite complex, not least because such materials have both civilian and military uses, thus complicating management across their lifecycles.
Source: Courtesy of C.K. Chou, Lawrence Livermore National Laboratory

should be as complete and transparent as possible, recognizing that, at the margin, military security concerns will undoubtedly arise.

2. Development and deployment of sensor and materials security systems globally that can help assure the integrity of nuclear material storage and management, and prevent theft or diversion into informal channels.

3. Sponsorship of regular technology transfer activities, whereby global nuclear operations, particularly nuclear power and fuel cycle activities, can all be raised to world class safety and risk reduction levels.

18.10 CONCLUSION

In many ways, the integration of environmental considerations and national security issues is fascinating from an industrial ecology perspective, both as a case study, and also for the complexity of the substantive issues that are raised. The nuclear materials case study, for example, with its dimensions of energy security, trade-offs with fossil fuel combustion, and military and terrorism implications, is an industrial ecology challenge in itself. Beyond that, the cultural and institutional implications of environment being treated as (literally) strategic provide a useful primer in the implementation of the principles of industrial ecology in the real world of policy and high politics.

REFERENCES

Allenby, B.R., T.J. Gilmartin, and R. Lehman, eds. *Environmental Threats and National Security.* Proceedings of a conference. Lawrence Livermore National Laboratory. In press.

Center for Strategic and International Studies (CSIS) Global Organized Crime Project. *The Nuclear Black Market.* 1996.

Department of Defense, Office of the Under Secretary of Defense, Acquisition and Technology. *Report of the Defense Science Board Task Force on Environmental Security.* 1995.

Department of Energy, Energy Information Agency. *International Energy Outlook.* 1996.

Executive Office of the President. *A national security strategy of engagement and enlargement.* 1996.

Gleick, P.H. "Water and conflict." *International Security* 18(1):79–112 (1993).

Goldstone, J.A. *Debate, Environmental Change and Security Project Report.* 2:66–71 (1996).

Homer-Dixon, T. "Environmental scarcities and violent conflict: evidence from cases." *International Security* 19(1):5–40. 1994.

Homer-Dixon, T. *Strategies for studying causation in complex ecological political systems.* AAAS/University of Toronto, 1995.

Homer-Dixon, T., J.H. Boutwell, and G.W. Rathjens. "Environmental change and violent conflict." *Scientific American* 38–45. 1993.

Liverman, D.M. and K.L. O'Brien, "Global warming and climate change in Mexico." *Global Environmental Change* 351–364. 1991.

Mathews, J.T. "Redefining security." *Foreign Affairs* 68:162–177. 1989.

Mathews, J.T. "Power shift." *Foreign Affairs* 76(1):50–66. 1997.

Morehouse, E.T. Jr. "Preventing pollution and seeking environmentally preferable alternatives in the U. S. Air Force," in B. R. Allenby and D. J. Richards, eds. *The Greening of Industrial Ecosystems.* Washington, DC, National Academy Press: 1994, 149–164.

National Science and Technology Council (NSTC). *National security science and technology strategy.* 1995.

Shcherbak, Y.M. "Ten years of the Chernobyl era: the environmental and health effects of nuclear power's greatest calamity will last for generations." *Scientific American* 44–49. 1996.

Turner, B.L. II, W.C. Clark, R.W. Kates, J.F. Richards, J.T. Mathews, and W.B. Meyer, eds. *The Earth as Transformed by Human Action.* Cambridge, UK, Cambridge University Press: 1990.

EXERCISES

1. Using the three part test given in this chapter, determine whether the following should be considered environmental security issues, and, if so, for whom. Justify your conclusions.

    a. As a result of shifts in global climate patterns, China suffers a series of droughts that begin to fuel significant internal and external population migrations.

    b. A combination of poor weather, infrastructure failure, and population growth lead to civil war in a central African nation.

    c. Economic and population growth leads to a precipitous drop in biodiversity throughout Asia.

    d. Global climate change leads to more severe storm patterns in Asia, with severe crop disruption throughout the Indian subcontinent.

2. You are the European negotiator seeking to achieve a lasting peace between Israel and its neighbors. The issue of water allocation has come up, with the Palestinians complaining that Israel is taking water from aquifers underlying their territory, leaving them with little but salinized and polluted groundwater. What confidence building measures would you suggest under these circumstances?

3. Using the S&T framework in Figure 18.3, develop a plan for monitoring the stability of the Indian subcontinent from an environmental security perspective. Suppose that a model indicates that an effect of global climate change over the next decade will be to dramatically reduce the monsoon rains over India. What effects would you expect that to have? Produce and defend a plan for collaborative confidence building measures to maintain the stability of that region in such an eventuality.

4. a. The relations between India and two of its neighbors, Pakistan and China, have traditionally been unsettled. Looking at a map of the Indian subcontinent, what potential environmental security issues can you identify from the viewpoint of i) India, and ii) the United States?

    b. China is an avowed nuclear power, and is suspected of helping Pakistan develop such a capability. India is suspected of having a similar capability. Given their increasing demand for energy, it is probable that each state will rely, at least to some extent, on nuclear power for the foreseeable future. Given the history of conflict in the region, what confidence building measures would you suggest?

5. You are a Mexican diplomat. What environmental security issues would you consider in evaluating your national security with regard to the United States? To Canada?

Index